Springer Proceedings in Mathematics & Statistics

Volume 487

This book series features volumes composed of selected contributions from workshops and conferences in all areas of current research in mathematics and statistics, including data science, operations research and optimization. In addition to an overall evaluation of the interest, scientific quality, and timeliness of each proposal at the hands of the publisher, individual contributions are all refereed to the high quality standards of leading journals in the field. Thus, this series provides the research community with well-edited, authoritative reports on developments in the most exciting areas of mathematical and statistical research today.

Rajesh Kumar Sharma ·
Shailesh Kumar Srivastava · Ali Cemal Benim
Editors

Mathematical Theory and Simulation of Scientific Problems

FIAM-2023, Dubai, UAE, December 21–22

Springer

Editors
Rajesh Kumar Sharma
Department of Mathematics
North Eastern Hill University
Shillong, Meghalaya, India

Shailesh Kumar Srivastava
Department of Mathematics
Sardar Vallabhbhai National Institute of Technology
Surat, Gujarat, India

Ali Cemal Benim
Faculty of Mechanical and Process Engineering
Duesseldorf University of Applied Sciences
Düsseldorf, Germany

ISSN 2194-1009 ISSN 2194-1017 (electronic)
Springer Proceedings in Mathematics & Statistics
ISBN 978-981-96-2578-9 ISBN 978-981-96-2579-6 (eBook)
https://doi.org/10.1007/978-981-96-2579-6

Mathematics Subject Classification: 37-06, 92-06, 76-06, 93-06, 00B25, 60-06

This Springer imprint is published by the registered company Springer Nature Singapore Pte Ltd.
The registered company address is: 152 Beach Road, #21-01/04 Gateway East, Singapore 189721, Singapore

Preface

We are pleased to present the proceedings of the 6th International Conference on Frontiers in Industrial and Applied Mathematics (FIAM-2023), held at the Department of General Science, Birla Institute of Technology and Science, Pilani-Dubai Campus, UAE, from 21 to 22 December 2023. The FIAM-2023 is the 6th sequel of the FIAM conference series, which was started in 2018 by a group of faculty members working in different institutes in India. The focus of this conference is on the applications and applied research in various fields of mathematics.

The scientific and technical program of the two-day international conference has invited talks and contributed oral presentations in the areas of computational and theoretical fluid dynamics, server queues, Lie group theory, fixed point theory, bio-mathematics, fuzzy, nonlinear dynamics, fractional calculus and approximation theory. These areas have numerous applications in the emerging technological and knowledge-based industries. A sound understanding of the same shall lead to the emergence of new ideas that can help to build trained professionals who can serve in the knowledge-based industries.

The papers presented at FIAM-2023 reflect the diversity and depth of current research in computational and applied mathematics. The selected papers after peer review are published in this proceedings. Most of the papers deal with mathematical theory and its applications to the various disciplines of engineering and sciences. Among the many high-quality papers, some key contributions stood out for their innovative approach and practical impact. For example, the study on chaotic Hamiltonian systems, by Alexandre R. Nieto, Jesus M. Seoane and Miguel A. F. Sanjuan, explored new methodologies in nonlinear dynamics. Another significant paper, authored by Deepika Parmar, B. V. Rathish Kumar, S. V. S. S. N. V. G. Krishna Murthy and Sumant Kumar, focused on the time-fractional derivative models applied to nanofluid dynamics in porous cavities. Additionally, a paper by C. Deniz Canal and A. C. Benim analyzed co-firing in swirl burners, providing vital insights into optimizing industrial combustion processes. In an another standout paper, Lucas Wangwe and Santosh Kumar explored a common fixed-point theorem for multivalued mappings in weak-partial *b*-metric space with a specific application that enhances the understanding of fixed-point theory and its mathematical implications. A paper by Vrushali

Khaladkar and Manish Kumar introduced a triple secure encryption scheme for three-channel images based on the Hankel transform, geometric transforms and hyper-chaotic maps, contributing significantly to secure data transmission. Additionally, noteworthy contributions include the work of Vora Hardagna Vatsal, Brajesh Kumar Jha and Tajinder Pal Singh, who presented a significant study on the effects of ER and Orai flux on fractional calcium diffusion in neuronal disorders, offering a mathematical perspective on complex biological processes. Sanjalee Maheshwari and Rajesh Sharma have contributed a numerical study on the thermo-convective instability of Au-Fe_3O_4 hybrid bi-viscous Bingham nanofluid. These papers, among others, reflect the breadth of research presented at the conference, showcasing the versatility of applied mathematics in addressing diverse challenges across industries.

This book illustrates the mathematical simulation of scientific problems and cutting-edge development in multiple branches of mathematics, including various computational and modeling techniques with case studies and concrete examples. It can be used also as a reference book which is especially suitable for graduate students and scholars who are interested in the real applications of mathematics, and professionals working in the interdisciplinary domains of mathematics and engineering. The target readers for this book include academicians, researchers, graduate students and industry experts interested in the latest advancements in mathematical techniques and their industrial applications.

We would like to take this opportunity to offer our thanks to all the contributors for their excellent work, and to reviewers who took time and effort to provide their valuable comments in time and help to improve the quality of the papers. Our high appreciation is to all invited speakers and all participants for their enthusiastic participation in this conference. We also thank Birla Institute of Technology and Science, Pilani-Dubai Campus, for hosting the event and providing a stimulating environment for fruitful discussions and collaborations. Our sincere and special thanks to the Springer Nature team for all the support offered during the journey to publish this book.

Ahmedabad, India — Rajesh Sharma
Düsseldorf, Germany — Ali Cemal Benim
Dubai, UAE — Maneesh
Tomsk, Russia — Mikhail A. Sheremet

Contents

Exploring Noisy Chaotic Hamiltonian Systems

Alexandre R. Nieto, Jesús M. Seoane, and Miguel A. F. Sanjuán

Abstract In this work, we investigate the impact of noise on two-degree-of-freedom Hamiltonian systems where chaotic scattering takes place. We show that noise influences the system differently based on whether the isopotential curves are open or closed. Within the open regime, noise induces a phenomenon of noise-enhanced trapping, leading to an increase in the average escape time for particular noise intensities. On the other hand, in the closed regime noise can generate escapes that are not possible in the conservative system.

Keywords Hamiltonian systems · Chaotic scattering · Noise

1 Introduction

Chaotic scattering in open Hamiltonian systems holds significant importance in nonlinear science, finding applications in classical [1, 2] and quantum physics [3, 4]. This phenomenon extends its relevance to diverse fields such as chemistry [5, 6] and biology [7, 8]. While much of the research in chaotic scattering has focused on purely conservative systems [9–12], many real-world situations are subject to environmental influences or internal irregularities. Consequently, recent studies have incorporated perturbations into conservative systems, aiming to model the system's interaction with its surroundings. Specifically, the impacts of dissipation [13, 14], periodic forcing [15, 16], and noise [17–20] have been explored.

Concerning the influence of noise, recent research has predominantly concentrated on area-preserving maps. In this context, noise has been identified as a factor capable of disrupting small scales of Kolmogorov-Arnold-Moser (KAM) islands,

A. R. Nieto · J. M. Seoane · M. A. F. Sanjuán (✉)
Nonlinear Dynamics, Chaos and Complex Systems Group, Departamento de Física, Universidad Rey Juan Carlos, Móstoles, Madrid, Spain
e-mail: miguel.sanjuan@urjc.es

R. K. Sharma et al. (eds.), *Mathematical Theory and Simulation of Scientific Problems*, Springer Proceedings in Mathematics & Statistics 487,
https://doi.org/10.1007/978-981-96-2579-6_1

leading to escapes prohibited in deterministic systems [17, 18, 21]. However, it's worth noting that noise can also induce trajectories outside KAM islands to enter inside them and exhibit transient regular motion [17, 19]. Furthermore, another interesting phenomenon known as noise-enhanced trapping has been reported [19]. When this phenomenon takes place, small noise intensities play a constructive role by reducing the escape rate of the particles. This phenomenon manifests in fully chaotic and in mixed-phase-space systems, albeit with distinct underlying mechanisms. In fully chaotic systems, the escape rate reduction results from the blurring of the measure of the exits. On the other hand, in mixed-phase-space systems trapping is enhanced as certain trajectories within the chaotic sea can enter regions which are defined by KAM islands in the absence of noise.

In this work, following the results reported in [22, 23], we explore the effects of noise in open an closed two-degree-of-freedom Hamiltonian systems, which in this context have received relatively less attention. We focus our analysis on the different response of the system depending on the existence of natural exits.

Within the open regime, we demonstrate the existence of an additional mechanism that explains the noise-enhanced trapping phenomenon in open Hamiltonian systems. As chaotic maps are often derived as Poincaré surfaces of section from continuous systems, the results might be expected to be similar to the ones reported in area-preserving maps. However, in open Hamiltonian systems, the introduction of noise in the velocity or in the spatial coordinates indirectly influences the energy. This crucial distinction gives rise to a novel and supplementary mechanism.

On the other hand, within the closed regime noise enhanced-trapping does not play a relevant role on the escape dynamics. Conversely, in this situation the introduction of noise transforms bounded Hamiltonian dynamics into a chaotic scattering scenario. We observed that the relationship between the average escape time and the intensity of noise adheres to distinct algebraic scaling laws for low and high noise intensities. This discrepancy arises due to the disruption of the stickiness [24, 25] associated with the KAM islands.

The organization of this manuscript is as follows. In Sect. 2, we describe our model, which is the Hénon-Heiles system incorporating additive uncorrelated Gaussian noise. Section 3 provides robust numerical evidence of noise-enhanced trapping within the open regime of the system. The noise-activated escape phenomenon in the closed regime is explored in Sect. 4. Finally, in Sect. 5, we present the main conclusions.

2 Model Description

To conduct research on the impact of noise in Hamiltonian systems, we used the widely studied Hénon-Heiles Hamiltonian [26] as our model. This system has garnered extensive attention in the nonlinear dynamics community due to its diverse dynamical behaviors [27–30]. The system features a nonlinear potential exhibiting triangular symmetry, defining a Hamiltonian of the form

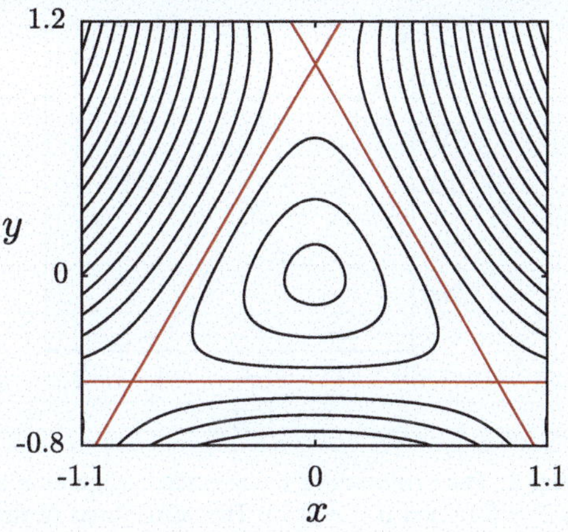

Fig. 1 Isopotential curves depicting various values of the potential of the Hénon-Heiles system. The energy value $E = E_e$ is represented by the red lines. As can be seen, energies $E \leq E_e$ generate bounded motion, while trajectories can escape trough three symmetrical exists when $E > E_e$

$$\mathcal{H} = \frac{1}{2}(\dot{x}^2 + \dot{y}^2) + \frac{1}{2}(x^2 + y^2) + x^2 y - \frac{1}{3}y^3, \tag{1}$$

where x and y represent spatial coordinates, and $\dot{x}$ and $\dot{y}$ denote the two components of the velocity.

The system experiences significant changes in its dynamical behavior at the energy threshold $E_e = 1/6$, known as escape energy. This threshold separates the open and closed regimes: for energies less than or equal to E_e, the isopotential curves are closed, and trajectories are bounded, while energies greater than E_e allow trajectories to escape. Fig. 1 illustrates isopotential curves for different energy values, providing a visual representation of the system's distinct regimes.

To consider the effect of noise, we have included a random force that we have modeled as uncorrelated additive Gaussian noise with zero mean. In this situation the equations of motion are

$$\begin{aligned} \ddot{x} &= -x - 2xy + \sqrt{2\xi}\,\eta_x(t), \\ \ddot{y} &= -y - x^2 + y^2 + \sqrt{2\xi}\,\eta_y(t), \end{aligned} \tag{2}$$

where $\eta_x(t)$ and $\eta_y(t)$ represent Gaussian white noise processes $X_t \sim \mathcal{N}(0, 2\xi)$ with intensity ξ and autocorrelation function $\langle \eta(t')\eta(t) \rangle = \sqrt{2\xi}\,\delta(t' - t)$.

3 Noise-Enhanced Trapping

We have characterized the dynamical behavior of the system by examining the average escape time of trajectories, denoted as $\bar{T}$. Over a broad range of noise intensities, we have computed average escape times for particles at energies $E = 0.18$ (in the mixed-phase space regime) and $E = 0.25$ (in the fully chaotic regime). The results,

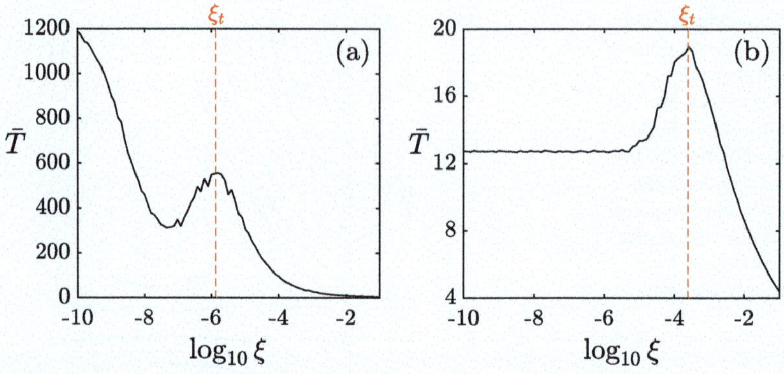

Fig. 2 The evolution of the average escape time as the noise intensity increases is depicted for **a** E = 0.18 and **b** E = 0.25. The vertical red dashed lines indicate the noise intensity, denoted as ξ_t, that induces noise-enhanced trapping

illustrated in Fig. 2, reveal either a relative or an absolute maximum in the average escape time, indicating that trapping is enhanced for specific noise values, ξ_t. Notably, strong noise intensities lead to a sharp decrease in the escape time, while weak noises generate different behaviors depending on the regime. In the fully chaotic regime (see Fig. 2b), the average escape time remains unchanged, converging to the value of the conservative system. Conversely, within the mixed-phase-space regime (see Fig. 2a), an initial decrease in $\bar{T}$ is observed. This decrease directly correlates with the reduction in stickiness of islands due to the effects of noise. While KAM islands technically do not exist in a noisy system, their remnants manifest through stickiness, retaining particles during extended transients. As noise intensity increases, stickiness diminishes, elevating the probability of escaping from a KAM region within a specific time.

The reduction in stickiness with increasing noise does not imply that KAM islands cease to influence the escape dynamics in noise-enhanced trapping. On the contrary, as demonstrated in prior research [19], particles following escaping orbits in the deterministic system may, due to noise effects, transition into KAM islands during long transients. Therefore, the initial decrease in average escape time is attributed to stickiness reduction, while noise-enhanced trapping is closely linked to KAM structures. As a matter of fact, noise induces fluctuations in energy, enabling particles to enter the structure of existing KAM islands for different energy values. However, this occurs only for specific noise intensities that align with the fine structure of the islands.

The noise-enhanced trapping phenomenon is generated by a small amount of "unusual" trajectories that reduce their energy due to the stochastic fluctuations and remain in the scattering region during astonishingly long transients. As an example, in Fig. 3 we show the evolution of the energy of one of these unusual trajectories. The escape time of the trajectory in the conservative system is $T = 12$, while in the presence of a noise intensity ξ_t the escape time increases until $T \approx 1400$. These kind of trajectories have a great effect on the average escape time.

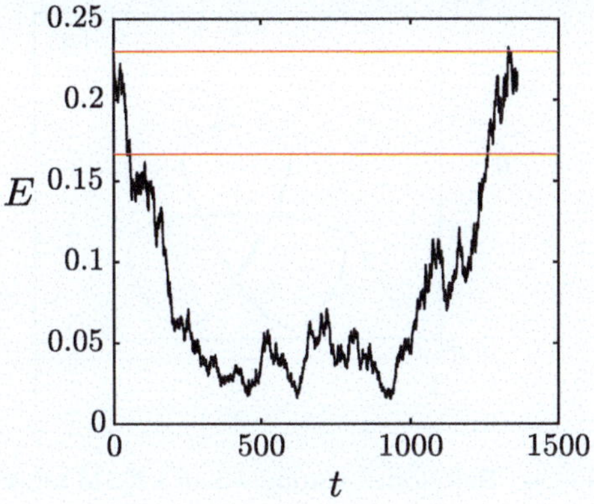

Fig. 3 Evolution of the energy of an unusual particle. The upper red line indicates the initial energy $E = 0.23$, while the lower refers to the escape energy $E_e = 1/6$. As can be observed, the trajectory moves during a long transient with very low energy $E < 1/6$. The curve concludes when the trajectory makes its escape

4 Noise-Activated Escape

As mentioned in Sect. 2, when $E \leq 1/6$ trajectories remain confined within the potential well. Despite this, the dynamical behavior is intricate, as both quasiperiodic and chaotic orbits coexist. Depending on the initial conditions, a trajectory may evolve differently, either following a regular orbit within a KAM torus or engaging in chaotic motion. The dynamics undergo a significant shift when we introduce the influence of noise on the system. In this scenario, the previously described conditions change dramatically. As noise impacts the particles' velocities, energy conservation is no longer upheld. The system experiences tremors due to fluctuations in energy, and a sequence of positive energy fluctuations propels the particle toward escape, even in the presence of very low noise intensity.

To visualize the noise-activated escape, we show in Fig. 4 two trajectories starting at an initial condition that belongs to a KAM torus of the noiseless system. As can be seen in Fig. 4a, in the absence of noise the particle describes a quasiperiodic orbit, while it escapes when noise is included (see Fig. 4b).

Upon examining Fig. 4, a counterintuitive outcome becomes apparent, since in panel (a) the isopotential curves are closed, whereas in panel (b) they are open. At first glance, one might assume that the simulations were conducted with different energy values, such as $E = 1/6$ in panel (a) and $E > 1/6$ in panel (b). However, the opening of the potential well is a consequence of the presence of noise. Despite the noise intensity having a zero mean, the energy undergoes changes, enabling particles to evolve with higher or lower energies than at $t = 0$. Therefore, if we represent the isopotential curves at the moment of escape, they will appear open.

We have provided a qualitative explanation of the noise-activated escape phenomenon. To quantitatively demonstrate the impact of noise on escape dynamics, we now calculate the average escape time $\bar{T}$ as a function of noise intensity. For numerical simulations, we have varied the noise intensity within a broad range. For each noise intensity, the escape time was computed for 3×10^5 random initial conditions to determine the average escape time. The results of this simulation are

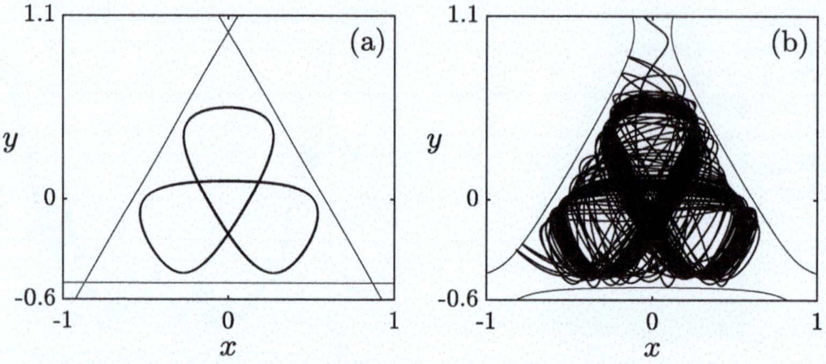

Fig. 4 Trajectories starting within a KAM torus are depicted in **a** the noiseless system and **b** in the presence of a noise intensity $\xi = 10^{-6}$. In both cases the initial energy is $E_0 = 1/6$. Even if the noiseless system is closed, the particle escapes after a finite time when noise is considered. The isopotential curves in panel (**b**) correspond to the energy of the system at the moment that the particle escapes ($E = 0.184$)

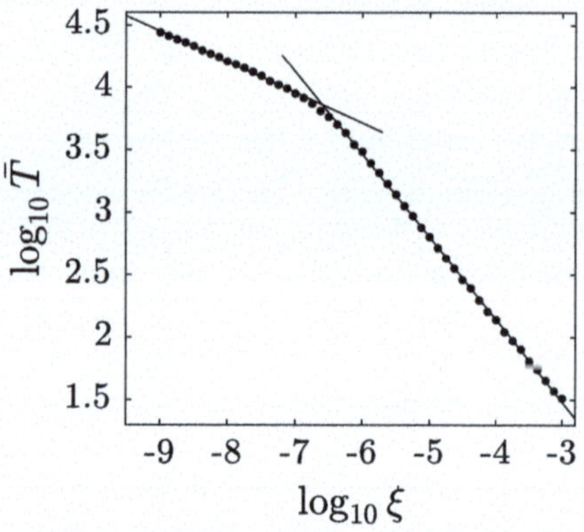

Fig. 5 Log-log plot of the evolution of the average escape time with increasing noise intensity. The solid dark lines correspond to two linear fits for low ($\log \bar{T}_l = -0.25 \log \xi + 5.16$) and high ($\log \bar{T}_h = -0.66 \log \xi - 1.15$) noise intensities

presented in Fig. 5, revealing two distinct regimes characterized by different algebraic laws. For low noise intensities, the average escape time follows an algebraic law $\bar{T}_l$, while high noise intensities are governed by another algebraic law $\bar{T}_h$. To illustrate the algebraic relationship between the variables, the results are depicted in a log-log plot accompanied by two linear fits. By examining the slope and the ln $\bar{T}$-intercept obtained through the least squares method, we can deduce the parameters of the underlying algebraic law

$$\bar{T} = \alpha \xi^{-\beta}, \tag{3}$$

where α and β are positive constants.

The reduction in the average escape time generated by the increase of the noise intensity can be evaluated by means of the derivative of Eq. (3). Giving the values of α and β, for low noise intensities we obtain

$$\left|\frac{d\bar{T}_l}{d\xi}\right| < \left|\frac{d\bar{T}_h}{d\xi}\right|. \tag{4}$$

This equation suggests that the algebraic law $\bar{T}_h$ fails to accurately predict the impact of noise on average escape time at low noise intensities, as it tends to overestimate the effect of noise on the escape time. Hence, it is evident that for low noise intensities there is a factor mitigating the decrease of $\bar{T}$. This factor is the stickiness of the KAM islands, which influences escape dynamics at low noise intensities, but becomes negligible at high ones [23].

5 Conclusions and Discussion

In summary, our investigation unveils distinct qualitative effects of noise in open and closed Hamiltonian systems. In open systems, a phenomenon known as noise-enhanced trapping emerges. Our findings indicate that this trapping is driven by a mechanism distinct from those reported in area-preserving maps. Specifically, it stems from the presence of unusual particles that undergo energy reduction due to stochastic fluctuations, resulting in prolonged transients.

In closed systems, noise primarily induces random energy fluctuations, enabling particles to surpass the threshold E_e and escape. The evolution of the average escape time displays a transition region that delineates two distinct regimes, each characterized by different algebraic scaling laws. Under the influence of noise, both open and closed Hamiltonian systems exhibit chaotic scattering. This suggests the possibility of modeling certain physical problems using noisy closed instead of open Hamiltonian systems. Given the different escape dynamics in each case, the choice of model depends on the system's behavior.

While our numerical evidence is based on the Hénon-Heiles Hamiltonian, we anticipate that similar phenomena may manifest in various Hamiltonian systems. The Barbanis potential [31], known for its applications in astrophysics [32, 33] and quantum dynamics [34], along with other Hamiltonians employed in modeling galactic movements [32, 35, 36], are plausible candidates to exhibit these phenomena.

References

1. Lai, Y.-C., Tél, T.: Transient Chaos: Complex Dynamics on Finite-Time Scales. Springer, New York (2010)
2. Zaslavsky, G.M.: The Physics of Chaos in Hamiltonian Systems. Imperial College Press, London (2007)
3. Lu, W.T., Sridhar, S., Zworski, M.: Fractal Weyl laws for chaotic open systems. Phys. Rev. Lett. **91**, 154101 (2003). https://doi.org/10.1103/PhysRevLett.91.154101
4. Stöckmann, H.-J.: Quantum Chaos: An Introduction. Cambridge University Press, Cambridge (1999)

5. Ezra, G.S., Waalkens, H., Wiggins, S.: Microcanonical rates, gap times, and phase space dividing surfaces. J. Chem. Phys. **130**, 164118 (2009). https://doi.org/10.1063/1.3119365
6. Kawai, S., Bandrauk, A.D., Jaffé, C., Bartsch, T., Palacián, J., Uzer, T.: Transition state theory for laser-driven reactions. J. Chem. Phys. **126**, 164306 (2007). https://doi.org/10.1063/1.2720841
7. Tél, T.: Chaotic advection, diffusion, and reactions in open flows. Chaos **10**, 89–98 (2000). https://doi.org/10.1063/1.166478
8. Schelin, A.B., Károlyi, G., de Moura, A.P.S., Booth, N.A., Grebogi C.: Fractal structures in stenoses and aneurysms in blood vessels. Phil. Trans. R. Soc. A **368**, 5605–5617 (2010). https://doi.org/10.1098/rsta.2010.0268
9. Bleher, S., Grebogi, C., Ott, E.: Bifurcation to chaotic scattering. Phys. D **46**, 87–121 (1990). https://doi.org/10.1016/0167-2789(90)90114-5
10. Contopoulos, G., Kandrup, H.E., Kaufmann, D.: Fractal properties of escape from a two-dimensional potential. Phys. D **64**, 310–321 (1993). https://doi.org/10.1016/0167-2789(93)90262-Y
11. Kandrup, H.E., Siopis, C., Contopoulos, G., Dvorak, R.: Diffusion and scaling in escapes from two-degrees-of-freedom Hamiltonian systems. Chaos **9**, 381–392 (1999). https://doi.org/10.1063/1.166415
12. Aguirre, J., Vallejo, J.C., Sanjuán, M.A.F.: Wada basins and chaotic invariant sets in the Hénon-Heiles system. Phys. Rev. E **64**, 066208 (2001). https://doi.org/10.1103/PhysRevE.64.066208
13. Motter, A.E., Lai, Y-C.: Dissipative chaotic scattering. Phys. Rev. E **65**, 015205(R) (2001). https://doi.org/10.1103/PhysRevE.65.015205
14. Burton, L.G., Dullin, H.R., Altmann, E.G.: Probabilistic description of dissipative chaotic scattering. Phys. Rev. E **108**, 054223 (2023). https://doi.org/10.1103/PhysRevE.108.054223
15. Blesa, F., Seoane, J.M., Barrio, R., Sanjuán, M.A.F.: Effects of periodic forcing in chaotic scattering. Phys. Rev. E **89**, 042909 (2014). https://doi.org/10.1103/PhysRevE.89.042909
16. Nieto, A.R., Seoane, J.M., Alvarellos, J.E., Sanjuán, M.A.F.: Resonant behavior and unpredictability in forced chaotic scattering. Phys. Rev. E **98**, 062206 (2018). https://doi.org/10.1103/PhysRevE.98.062206
17. Rodrigues, C.S., Moura, A.P.S., Grebogi, C.: Random fluctuation leads to forbidden escape of particles. Phys. Rev. E **82**, 026211 (2010). https://doi.org/10.1103/PhysRevE.82.026211
18. da Silva, R.M., Manchein, C., Beims, M.W.: Exploring conservative islands using correlated and uncorrelated noise. Phys. Rev. E **97**, 022219 (2018). https://doi.org/10.1103/PhysRevE.97.022219
19. Altmann, E.G., Endler, A.: Noise-enhanced trapping in chaotic scattering. Phys. Rev. Lett. **105**, 244102 (2010). https://doi.org/10.1103/PhysRevLett.105.244102
20. Nieto, A.R. Seoane, J.M., Sanjuán, M.A.F.: Final state sensitivity in noisy chaotic scattering. Chaos Solitons Fractals **150**, 111181 (2021). https://doi.org/10.1016/j.chaos.2021.111181
21. Mills, P.: The influence of noise on a classical chaotic scatterer. Commun. Nonlinear Sci. Numer. Simulat. **11**, 899–906 (2006). https://doi.org/10.1016/j.cnsns.2005.02.003
22. Nieto, A.R. Seoane, J.M., Sanjuán, M.A.F.: Trapping enhanced by noise in nonhyperbolic and hyperbolic chaotic scattering. Commun. Nonlinear Sci. Numer. Simulat. **102**, 105905 (2021). https://doi.org/10.1016/j.cnsns.2021.105905
23. Nieto, A.R. Seoane, J.M., Sanjuán, M.A.F.: Noise activates escapes in closed Hamiltonian systems. Commun. Nonlinear Sci. Numer. Simulat. **105**, 106074 (2022). https://doi.org/10.1016/j.cnsns.2021.106074
24. Altmann, E.G., Motter, A.E., Kantz, H.: Stickiness in Hamiltonian systems: from sharply divided to hierarchical phase space. Phys. Rev. E **73**, 026207 (2006). https://doi.org/10.1103/PhysRevE.73.026207
25. Bunimovich, L.A., Vela-Arevalo, L.V.: Many faces of stickiness in Hamiltonian systems. Chaos **22**, 026103 (2012). https://doi.org/10.1063/1.3692974
26. Hénon, M., Heiles, C.: The applicability of the third integral of motion: some numerical experiments. Astron. J. **69**, 73–79 (1964). https://doi.org/10.1086/109234

27. Barrio, R., Blesa, F., Serrano, S.: Fractal structures in the Hénon-Heiles Hamiltonian. EPL **82**, 10003 (2008). https://doi.org/10.1209/0295-5075/82/10003
28. Barrio, R., Wilczak, D.: Distribution of stable islands within chaotic areas in the non-hyperbolic and hyperbolic regimes in the Hénon-Heiles system. Nonlinear Dyn. **102**, 403–416 (2020). https://doi.org/10.1007/s11071-020-05930-x
29. Zotos, E.E.: Classifying orbits in the classical Hénon-Heiles Hamiltonian system. Nonlinear Dyn. **79**, 1665–1677 (2015). https://doi.org/10.1007/s11071-014-1766-6
30. Nieto, A.R., Seoane, J.M., Sanjuán, M.A.F.: Period-doubling bifurcations and islets of stability in two-degree-of-freedom Hamiltonian systems. Phys. Rev. E **107**, 054215 (2023). https://doi.org/10.1103/PhysRevE.107.054215
31. Barrio, R., Blesa, F., Serrano, S.: Bifurcations and safe regions in open Hamiltonians. New J. Phys. **11**, 053004 (2009). https://doi.org/10.1088/1367-2630/11/5/053004
32. Contopoulos, G.: Order and Chaos in Dynamical Astronomy. Springer, Berlin (2002)
33. Navarro, J.F.: On the escape from potentials with two exit channels. Sci. Rep. **9**, 13174 (2019). https://doi.org/10.1038/s41598-019-49765-y
34. Babyuk, D., Wyatt, R.E., Frederick, J.H.: Hydrodynamic analysis of dynamical tunneling. J. Chem. Phys. **119**, 6482–6482 (2003). https://doi.org/10.1063/1.1605385
35. Kandrup, H.E., Novotny, S.J.: Phase mixing in unperturbed and perturbed Hamiltonian systems. Celest. Mech. Dyn. Astron. **88**, 1–35 (2004). https://doi.org/10.1023/B:CELE.0000009380.17257.98
36. de Assis, S.C., Terra, M.O.: Escape dynamics and fractal basin boundaries in the planar Earth-Moon system. Celest. Mech. Dyn. Astr. **120**, 105–130 (2014). https://doi.org/10.1007/s10569-014-9567-2

Numerical Study of Time-Fractional Double Diffusive Convective Flow in a Wavy Porous Cavity

Deepika Parmar, B. V. Rathish Kumar, and S. V. S. S. N. V. G. Krishna Murthy

Abstract The current numerical investigation illustrates how the fractional order parameter influences the dynamics of double-diffusive convective processes within a square porous enclosure featuring wavy walls. Specifically, the enclosure's left (hot) and right (cold) wavy walls are kept at constant temperatures and solute concentrations, while the remaining two walls are thermally insulated and impermeable to solute. To analyze the transient behavior of fluid flow, the Caputo time-fractional derivative is applied to the energy and mass transfer equations, and momentum transport equation is modeled using the Darcy approach. In addition, the $L1$-scheme is employed to estimate the fractional time-derivative term, and the entire mathematical model is subsequently solved using the Galerkin finite element method. The investigation encompasses various parameters such as the Rayleigh number (Ra), buoyancy ratio (N), Lewis number (Le), and fractional-order parameter (α). Moreover, simulations are carried out by varying the fractional-order parameter within the range ($0 < \alpha \leq 1$). The findings are presented through contour plots illustrating variations in isotherms, streamlines, and isoconcentrations, alongside numerical variations of the mean Nusselt number (Nu_m) and mean Sherwood number (Sh_m). It is evident from the results that the fractional parameter α significantly influences heat and solute transport phenomena, as well as the initial evolution states of streamlines, isotherms, and isoconcentrations.

Keywords Caputo fractional derivative · Porous media · Double diffusion · Nanofluids · Penalty finite element method

D. Parmar · S. V. S. S. N. V. G. Krishna Murthy (✉)
Department of Applied Mathematics, Defence Institute of Advanced Technology, Girinagar, Pune, India
e-mail: skmurthy@diat.ac.in

B. V. Rathish Kumar
Department of Mathematics and Statistics, Indian Institute of Technology Kanpur, Kanpur, India
e-mail: bvrk@iitk.ac.in

R. K. Sharma et al. (eds.), *Mathematical Theory and Simulation of Scientific Problems*, Springer Proceedings in Mathematics & Statistics 487,
https://doi.org/10.1007/978-981-96-2579-6_2

Nomenculture

x, y	Cartesian co-ordinates;
X, Y	Dimensionless co-ordinates;
u, v	Dimensional velocity components in x and y directions respectively;
U, V	Dimensionless velocity components in X and Y directions respectively;
t	Dimensional time;
T	Dimensional temperature;
P	Dimensional pressure;
k	Thermal conductivity;
L	Characteristic length;
r	Amplitude of wavy wall;
K	Permeability;
C'	Solute concentration;
C	Solute concentration, dimensionless;

Non-dimensional Parameters

Ra	Rayleigh number;
Le	Lewis number;
Nu_m	Mean Nusselt number;
Sh_m	Mean Sherwood number;
N	Buoyancy ratio;

Greek Symbols

α	Order of fractional derivative;
α_{nf}, α_{bf}	Thermal diffusivity of nanofluid and base fluid respectively;
β_T	Volumetric expansion coefficient with temperature;
$\beta_{C'}$	Volumetric expansion coefficient with concentration;
ρ	Density;
μ	Dynamic viscosity;
Ψ	Dimensionless stream function;
θ	Dimensionless temperature;
τ	Dimensionless time;

Subscripts

nf	Nanofluid;
bf	Base fluid;
s	Solid matrix;
c	Cold;
h	Hot.

1 Introduction

Double diffusion phenomena primarily arise when convection induces density fluctuations due to the transport of heat and solutes. This phenomenon is prevalent across various fields of science and engineering, including seawater flow, chemical reactions, nuclear waste movement, pollutant dispersion, and the metal manufacturing process [1–3]. Recent research has been focused on enhancing heat and mass transfer processes by employing porous media saturated with nanofluids and hybrid nanofluids [4]. Additionally, the use of different types of enclosures is crucial for improving heat and mass transport efficiency [5–7]. Furthermore, numerical simulations have been conducted for both Darcy [8] and non-Darcy [9] models to analyze double-diffusion phenomena.

Fractional-order models are increasingly used in modern science and engineering due to their ability to capture memory effects and non-local behavior, which are not present in traditional models. Fractional differential equations (FDEs) have proven to be effective tools for studying heat transfer and fluid flow processes in porous. Utilizing FDEs enables a more precise representation of heat transfer and fluid flow phenomena, effectively capturing the atypical behaviors observed in porous media systems that are often missed by traditional mathematical models [10, 11]. Karani et al. [12] investigated the initiation of thermal convection utilizing fractional order derivatives, demonstrating that fractional order models offer significantly higher accuracy in predicting the onset of thermal convection compared to classical models. Ahmed [13] employed the Caputo fractional derivative to investigate convective flow in an inclined wavy vented cavity filled with porous media. Consequently, considering the potential of fractional order models, this research aims to examine the impact of the Caputo fractional derivative on double diffusive phenomena.

2 Problem Description

Figure 1 depicts the schematic of working domain which is essentially a square with wavy vertical walls. The enclosure's left wall is set to higher temperature (T_h) and solute concentration (C_h'). The right wall is set to comparatively lower temperature (T_c) and solute concentration. (C_c'). The top and bottom walls are insulated and

Fig. 1 Physical domain

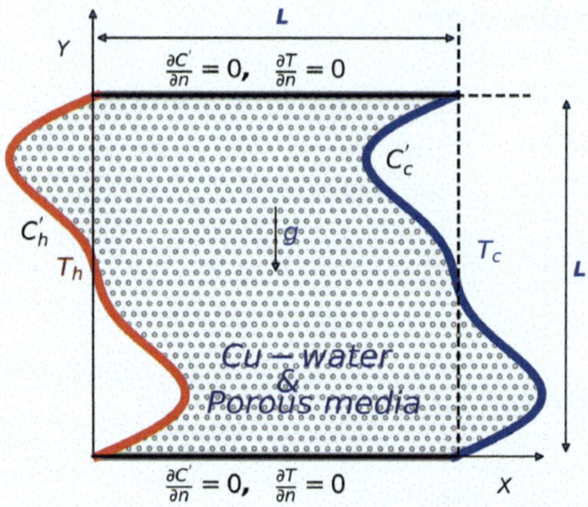

non-diffusive. Cu-water nanofluid saturates the porous enclosure. The working fluid undergoes laminar flow and is assumed to be Newtonian and incompressible. The solid matrix and working fluid maintain local thermal equilibrium conditions.

2.1 Governing Equations

The governing equations in dimensional form are given as [14]

Continuity Equation

$$\frac{\partial u}{\partial x} + \frac{\partial v}{\partial y} = 0 \tag{1}$$

Momentum Equation

$$\frac{\partial u}{\partial y} - \frac{\partial v}{\partial x} = -gK\frac{\rho_{nf}}{\mu_{nf}}\left[(\beta_T)_{nf}\frac{\partial T}{\partial x} + (\beta_{C'})_{nf}\frac{\partial C'}{\partial x}\right] \tag{2}$$

Energy Equation

$$\sigma\,(t_c)^{(a-1)}\frac{\partial^\alpha T}{\partial t^\alpha} + \left(u\frac{\partial T}{\partial x} + v\frac{\partial T}{\partial y}\right) = \alpha_{nf}\left(\frac{\partial^2 T}{\partial x^2} + \frac{\partial^2 T}{\partial y^2}\right) \tag{3}$$

Mass Transfer Equation

$$\epsilon\,(t_c)^{(a-1)}\frac{\partial^\alpha C'}{\partial t^\alpha} + \left(u\frac{\partial C'}{\partial x} + v\frac{\partial C'}{\partial y}\right) = D_m\left(\frac{\partial^2 C'}{\partial x^2} + \frac{\partial^2 C'}{\partial y^2}\right) \tag{4}$$

where x and y denote dimensional coordinates, and u and v are the dimensional velocity components in x and y directions, respectively. K, T, C', and t represent the permeability, dimensional temperature, solute concentration, and time, respectively.

2.2 Non-dimensional Governing Equations

The governing equations in non-dimensional form are presented for two-dimensional geometry

Momentum Equation

$$\frac{\partial^2 \Psi}{\partial X^2} + \frac{\partial^2 \Psi}{\partial Y^2} = -\frac{\mu_{bf}}{\mu_{nf}} \frac{(\rho\beta_T)_{nf}}{(\rho\beta_T)_{bf}} Ra \left[\frac{\partial \theta}{\partial X} + N \frac{\partial C}{\partial X} \right] \quad (5)$$

Energy Equation

$$\sigma \left(\tau_c\right)^{(\alpha-1)} \frac{\partial^\alpha \theta}{\partial \tau^\alpha} + \left(\frac{\partial \Psi}{\partial Y} \frac{\partial \theta}{\partial X} - \frac{\partial \Psi}{\partial X} \frac{\partial \theta}{\partial Y} \right) = \frac{\alpha_{nf}}{\alpha_{bf}} \left(\frac{\partial^2 \theta}{\partial X^2} + \frac{\partial^2 \theta}{\partial Y^2} \right) \quad (6)$$

Mass Transfer Equation

$$\epsilon \left(\tau_c\right)^{(\alpha-1)} \frac{\partial^\alpha C}{\partial \tau^\alpha} + \left(\frac{\partial \Psi}{\partial Y} \frac{\partial C}{\partial X} - \frac{\partial \Psi}{\partial X} \frac{\partial C}{\partial Y} \right) = \frac{1}{Le} \left(\frac{\partial^2 C}{\partial X^2} + \frac{\partial^2 C}{\partial Y^2} \right) \quad (7)$$

Here, X and Y represent the non-dimensional coordinate components, while Ψ, θ, C, and τ correspond to the dimensionless streamlines, temperature, concentration, and time, respectively. Additionally, ρ, μ, ϵ, and σ stand for density, dynamic viscosity, porosity, and ratio of heat capacities. The subscript nf is used to indicate nanofluid, and bf refers to the base fluid. The terms Ra and Le correspond to the Rayleigh and Lewis numbers. The following non-dimensional variables are used to present the non-dimensional governing equations.

$$\begin{aligned} &X = x/L, \;\; Y = y/L, \;\; U = \frac{uL}{\alpha_{bf}} = \frac{\partial \Psi}{\partial Y}, \;\; V = \frac{vL}{\alpha_{bf}} = -\frac{\partial \Psi}{\partial X}, \;\; \tau = \frac{t\alpha_{bf}}{L^2} \\ &\theta = \frac{T - T_c}{T_h - T_c}, \;\; Ra = \frac{gK\rho_{bf}(\beta_T)_{bf} L (T_h - T_c)}{\mu_{bf}\alpha_{bf}}, \;\; C = \frac{C' - C'_c}{C'_h - C'_c} \\ &Le = \frac{\alpha_{bf}}{D_m}, \;\; N = \frac{(\beta_{C'})_{nf}}{(\beta_T)_{nf}} \frac{(C'_h - C'_c)}{(T_h - T_c)} \end{aligned} \quad (8)$$

The associated boundary conditions

$$\begin{aligned}
&\Psi = 0,\ \frac{\partial \theta}{\partial n} = 0,\ \frac{\partial C}{\partial n} = 0,\ \text{on top and bottom walls,}\\
&\Psi = 0,\ \theta = 1,\ C = 1,\ \text{on left wall,}\\
&\Psi = 0,\ \theta = 0,\ C = 0,\ \text{on right wall.}
\end{aligned} \tag{9}$$

3 Solution Methodology, Grid Independence, and Model Validation

The governing equations (1)–(3) are solved utilizing the Galerkin finite element method [15] and the Caputo derivative term is discretized employing the L1-scheme [16]. A grid independence test is shown in Table 1 and conducted, employing various grid types denoted as $G1$ through $G8$, with total element counts of 13240, 13790, 14100, 14640, 14978, 15628, 15880, and 17490 respectively. The mean Nusselt number, mean Sherwood number, and relative percentage error values are calculated. Notably, it is observed that grid $G6$ yields optimal results for our computations. Moreover, the model's validation is conducted against previously published findings [14] in the literature for $Ra = 100$, $N = 0$, and $Le = 10$, as illustrated in Fig. 2.

Table 1 Grid Independence test is performed at $\alpha = 1$, $Ra = 10^2$, $Le = 10$, $\epsilon = 0.4$, and $N = 0$

Grid type	No. of elements	Nu_m	Relative % error	Sh_m	Relative % error
G1	13240	0.871997		0.927048	
G2	13790	0.875510	0.40%	0.929471	0.26%
G3	14100	0.873635	0.21%	0.928299	0.12%
G4	14640	0.875739	0.24%	0.929614	0.14%
G5	14978	0.874369	0.15%	0.928662	0.10%
G6	15628	0.873744	0.07%	0.928295	0.03%
G7	15880	0.873105	0.07%	0.928904	0.06%
G8	17490	0.873276	0.02%	0.929088	0.02%

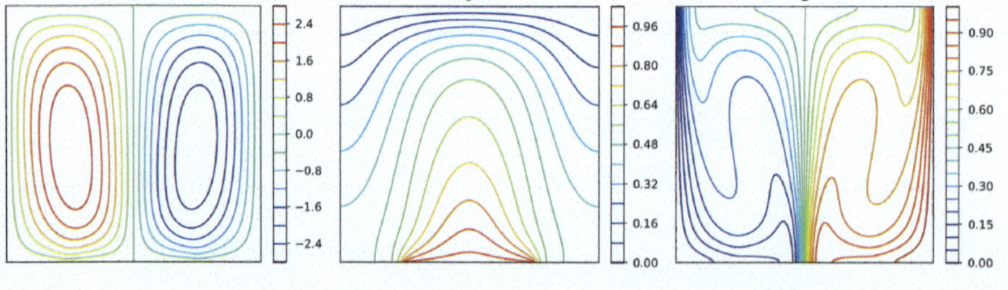

Fig. 2 Comparison of present results with the existing results in the literature [14]

4 Results and Discussions

The extensive findings of this study are depicted through the contour variations in streamlines, isotherms, and isoconcentration, along with plots illustrating the numerical changes in the mean Nusselt number and mean Sherwood number within specific parameter ranges. Throughout the study, values of parameters $\sigma = 1$ and $\epsilon = 0.4$ are fixed.

The evolution period of streamlines, isotherms, and isoconcentration for $\alpha = 0.95$ at $Ra = 100$, $Le = 10$, and $N = 1$ are illustrated in Fig. 3. Initially, when $\tau = 0.01$, it is noted that the streamline circulations densely cluster around the hot wall, forming a vertical elliptical zone close to the left wall. Since the fluid and solute movement is primarily concentrated near the left wall, the thermal and concentration layers are confined to a small region close to the hot wall. Further, as time incenses the streamline circulations, thermal and solute layers moving toward right wall due to the fluid and solute transport. By $\tau = 1.5$, fully developed flow is achieved, with no further changes occurring thereafter.

Figure 4 illustrates the variations in the Nu_m and Sh_m for different values of α, including $\alpha = 1, 0.95, 0.85, 0.75,$ and 0.65, at $\tau = 0.01, 0.05, 0.1, 0.5, 1, 1.5,$ and 2. It is observed that the Nu_m and Sh_m values are initially very high for each α value. However, as time progresses, these values tend to decrease, with smaller α values exhibiting faster convergence to the steady-state condition compared to larger α values.

5 Conclusion

The main goal of this study is to demonstrate the influence of the time-fractional derivative on the dynamics of heat and solute transport processes. To achieve this goal, a square porous enclosure with a wavy vertical wall is utilized as the experimental setup. The analysis incorporates the Darcy model, while the Caputo time derivative is applied to the energy and mass transfer equations. A key finding of this investigation is the substantial variation in the evolution patterns of streamlines, isotherms, and isoconcentrations across different α values. Furthermore, it is observed that the time needed to reach steady-state conditions is considerably extended for smaller α values, while it is relatively shorter for larger α values.

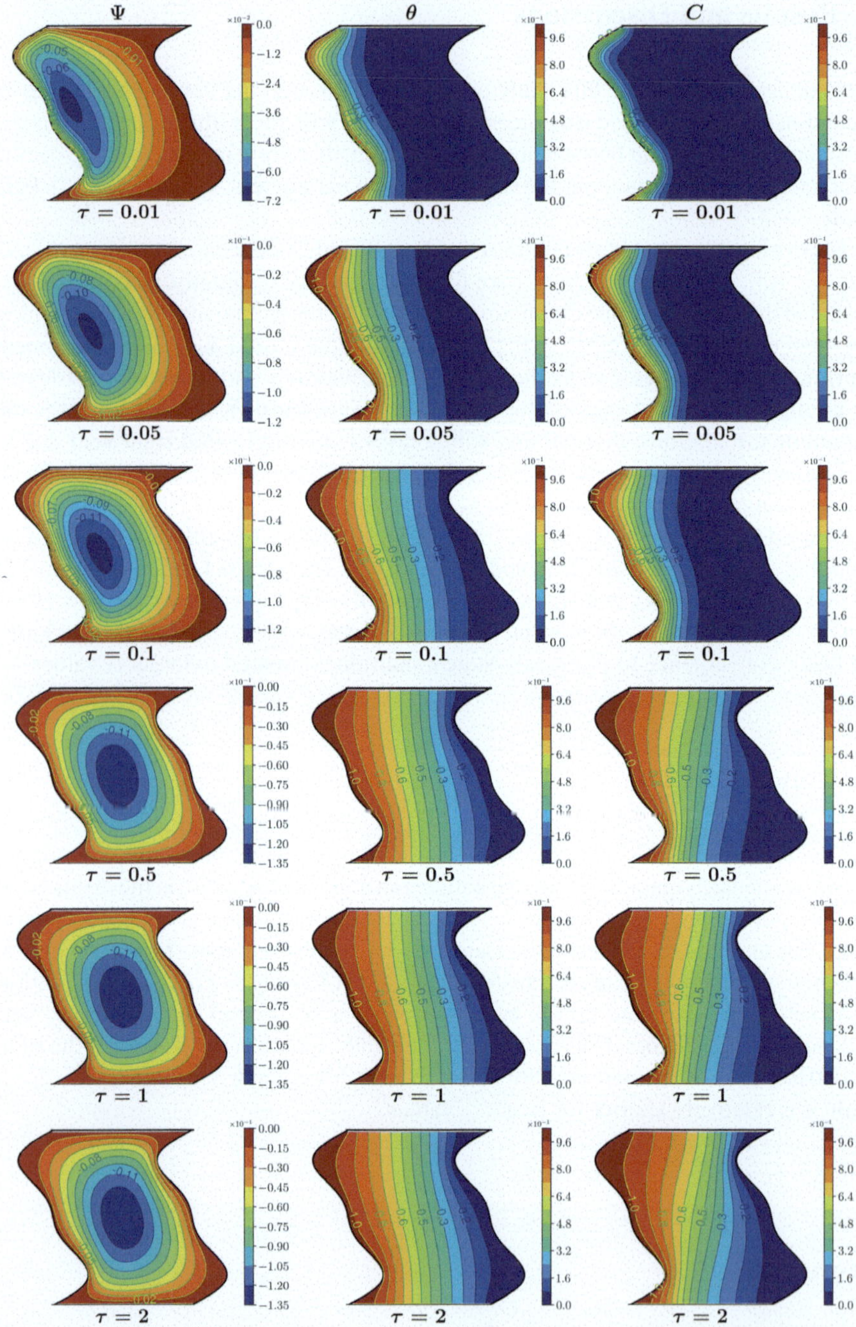

Fig. 3 Evolution state of streamlines, isotherms, and concentrations contours for $\alpha = 0.95$ at $Ra = 10^2$, $Le = 10$, $\epsilon = 0.4$, and $N = 1$

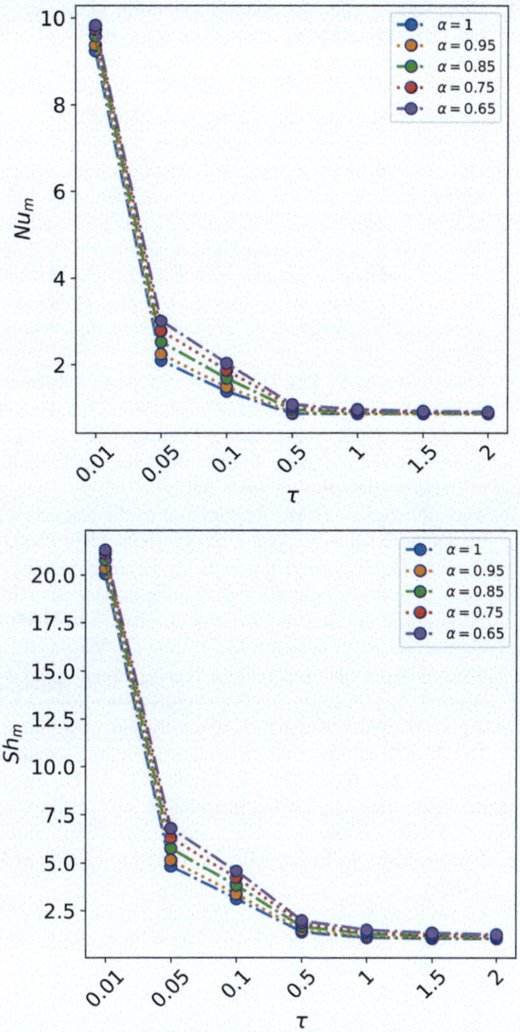

Fig. 4 Nu_m and Sh_m plots for various α values at different time stages

Acknowledgements The authors are thankful to the SPARC Project **(Ref No. SPARC/2019-2020/P2465/SL Dt.: 28.07.2023)**, funded by the Ministry of HRD, Govt. India, and steered by IIT-KGP. Additionally, the first author is grateful to DIAT for providing the financial support to attend FIAM-2023.

References

1. Huppert, H.E., Moore, D.R.: Nonlinear double-diffusive convection. J. Fluid Mech. **78**(4), 821–854 (1976)
2. Huppert, H.E., Sparks, R.S.J.: Double-diffusive convection due to crystallization in magmas. Ann. Rev. Earth Planet. Sci. **12**(1), 11–37 (1984)

3. Nield, D.A., Kuznetsov, A.V.: The onset of double-diffusive convection in a nanofluid layer. Int. J. Heat Fluid Flow **32**(4), 771–776 (2011)
4. S. Kumar, Rathish Kumar, B.V., Krishna Murthy, S.V.S.S.N.V.G., Parmar, D.: Double-diffusive convective flow of hybrid nanofluid in an inverted t-shaped porous enclosure: a numerical study. Numer. Heat Transf. Part A Appl. 1–25 (2023)
5. Fattahi, A., Hajialigol, N., Delpisheh, M., Karimi, N.: Lattice-boltzmann numerical simulation of double-diffusive natural convection and entropy generation in an n-shaped partially heated storage tank. Eng. Anal. Bound. Elem. **146**, 105–118 (2023)
6. Kumar, S.., Rathish Kumar, B.V., Krishna Murthy, S.V.S.S.N.V.G.: Double diffusive convective flow study of a hybrid nanofluid in an inverted t-shaped porous enclosure under the influence of soret and dufour prameters. ASME J. Heat Mass Transf. **145**(10), 102501 (2023)
7. Khan, Z.H., Khan, W.A., Qasim, M., Du, M.: Double-diffusive flow in a porous right-angle trapezoidal enclosure with constant heat flux. Math. Methods Appl. Sci. **45**(6), 3305–3317 (2022)
8. Malashetty, M.S., Pal, D., Kollur, P.: Double-diffusive convection in a darcy porous medium saturated with a couple-stress fluid. Fluid Dyn. Res. **42**(3), 035502 (2010)
9. Kumar, S., Gangawane, K.M.: Double-diffusive convection in a rectangular cavity subjected to an external magnetic field with heated rectangular blockage insertion for liquid sodium–potassium alloy. Phys. Fluids **34**(2) (2022)
10. Zhou, L., Selim, H.M.: Application of the fractional advection-dispersion equation in porous media. Soil Sci. Soc. Am. J. **67**(4), 1079–1084 (2003)
11. Parmar, D., Rathish Kumar, B.V., Krishna Murthy, S.V.S.S.N.V.G., Kumar, S.: Numerical study of entropy generation in magneto-convective flow of nanofluid in porous enclosure using fractional order non-darcian model. Phys. Fluids **35**(9) (2023)
12. Karani, H., Rashtbehesht, M., Huber, C., Magin, R.L.: Onset of fractional-order thermal convection in porous media. Phys. Rev. E **96**(6), 063105 (2017)
13. Ahmed, S.E.: Caputo fractional convective flow in an inclined wavy vented cavity filled with a porous medium using al2o3-cu hybrid nanofluids. Int. Commun. Heat Mass Transf. **116**, 104690 (2020)
14. Bourich, M., Hasnaoui, M., Amahmid, A.: Double-diffusive natural convection in a porous enclosure partially heated from below and differentially salted. Int. J. Heat Fluid Flow **25**(6), 1034–1046 (2004)
15. Reddy, J.N.: An Introduction to the Finite Element Method, vol. 3. McGraw-Hill New York (2013)
16. Li, D., Liao, H.-L., Sun, W., Wang, J., Zhang, J.: Analysis of l1-galerkin fems for time-fractional nonlinear parabolic problems. Commun. Comput. Phys. **24**(1), 86–103 (2018)

Numerical Study of Co-Firing in Swirl Burner Using Coal-Biomass Blends

C. Deniz Canal and A. C. Benim

Abstract The co-firing of biomass with coal presents an appealing approach for reducing CO_2 emissions and leveraging renewable energy sources. Oxy-fuel combustion, employing pure oxygen instead of air, stands out as a noteworthy technology for efficient CO_2 capture. This study focuses on the numerical investigation of the flame resulting from the co-combustion of pulverized coal and biomass. Utilizing the Fluent commercial CFD code, 2D numerical co-combustion calculations are conducted in a 100 kW vertical down-fired boiler equipped with a swirl burner. The accuracy of the pulverized coal combustion model is verified against experimental data before extending the model to incorporate biomass co-combustion. RANS turbulence models are employed to simulate the swirl burner, determining the most suitable model for co-combustion. The study explores three different blending ratios, such as 25%Biomass-75%Coal, 50%Biomass-50%Coal and 75%Biomass-25%Coal under an oxy-fuel atmosphere. Additionally, the investigation delves into the impact of turbulent particle dispersion on co-combustion flame.

Keywords Biomass combustion · Co-firing · Pulverized fuel combustion

1 Introduction

In recent decades, the climate has undergone changes attributed to the rise in greenhouse gases. Notably, CO_2 is a significant contributor to the greenhouse gas effect. Simultaneously, the demand for energy is on a constant upswing, even as traditional energy sources become more limited. Globally, combustion processes have served as a pivotal means to fulfil escalating energy requirements. Coal, as a fossil fuel, remains a prevalent choice for fuel in these processes. However, recognizing environmental

C. D. Canal (✉) · A. C. Benim
Center of Flow Simulation, Düsseldorf University of Applied Sciences, Düsseldorf, Germany
e-mail: cansu.denizcanal@hs-duesseldorf.com

A. C. Benim
e-mail: alicemal@prof-benim.com

R. K. Sharma et al. (eds.), *Mathematical Theory and Simulation of Scientific Problems*, Springer Proceedings in Mathematics & Statistics 487,
https://doi.org/10.1007/978-981-96-2579-6_3

concerns, there has been a growing shift toward renewable energy sources to address energy needs. Combustion, therefore, continues to play a crucial role, particularly in the context of renewable energies. Biomass, as a renewable energy source, relies on combustion processes for its conversion into usable energy [1–4].

Blending biomass with coal presents an attractive prospect for reducing CO_2 emissions and promoting the utilization of renewable energy sources. An additional benefit of this approach is its cost-effectiveness when implemented in existing coal-fired power plants. Moreover, co-firing stands out as a straightforward method for incorporating biomass into energy production, providing a simpler alternative to using fossil fuels. The properties of the fuel are crucial factors influencing the combustion process, and the ratio of fuel blends significantly impacts the outcomes of co-firing. Notably, coal exhibits a higher sulphur content compared to biomass, introducing variations in the emission gases content following the co-firing process [5–7].

Three co-combustion methods are distinguished: direct co-firing, indirect co-firing, and parallel co-firing. Direct co-firing is advantageous in terms of cost and remains the most widely employed method for co-combustion. The second method involves the gasification of biomass before co-combustion. However, the need to cool down the gases post-gasification results in elevated operational costs. Parallel co-firing, the third option, entails the separate combustion of coal and biomass. This approach is commonly utilized, particularly in the context of paper industries [6].

The combustion of a solid fuel particle involves four distinct stages: evaporation, devolatilization, volatile combustion, and char combustion. Initially, the particle undergoes evaporation, during which water is released. In the devolatilization stage, volatile matter and tar are emitted from the fuel particle. The volatile gases undergo combustion with oxygen in the volatile combustion step. Finally, in the char combustion stage, char reacts with oxygen, producing CO and CO_2. These stages may occur simultaneously or progressively, depending on factors such as particle size and heating rate. The composition of char, gas, and tar is influenced by various parameters, including elemental analysis, particle size, and boiler temperature [7].

This research delves into the numerical investigation of coal and biomass co-firing in a 100-kW pulverized boiler. The modelling is conducted using the Fluent commercial CFD code. The accuracy of the pulverized coal combustion model is established through validation against experimental data. Subsequently, the validated model is applied to model the co-combustion flame. Three distinct blending ratios are explored, such as 75–25% biomass-coal, 50–50% biomass-coal, 25–75% biomass-coal within an oxy-fuel atmosphere. Additionally, the study examines the impact of turbulent particle dispersion on the co-combustion flame. The outcomes of the combustion model are presented, highlighting the influence of turbulence on a lab-scale burner, temperature distribution, and gaseous emissions across various biomass ratios.

2 Numerical Modelling

Although alternative numerical procedures such as the finite element method [8] are also used, the most commonly applied discretization procedure in computational fluid dynamics has been the finite volume method [9], which is also applied in the present study. The computational fluid dynamics (CFD) code employed in this study is Ansys Fluent 18.0 [10], utilizing a finite volume method for discretization. A coupled solver is applied, and the convection terms are discretized using a second-order upwind scheme. Gradient calculations are performed using the least squares cell-based method [11–13]. For this case, the Eulerian–Lagrangian approach is adopted [14, 15]. The gas phase is treated as a continuum, with the Navier–Stokes (N-S) equations being solved [16, 17]. Simultaneously, the particle phase is addressed by tracking the movement of particles through the calculated flow field, modelling the turbulent dispersion by the so-called discrete random walk model [18]. Every 30th iteration involves the injection of particles into the boiler. The particles are assumed to possess a spherical shape. Coal and biomass particle size distribution are modelled using the Rosin–Rammler distribution [19, 20].

The RANS (Reynolds-Averaged Navier–Stokes) method [21, 22] is employed to model gas turbulence. It is known that the computational modelling of turbulent swirling flows can be very demanding [23, 24]. In a prior study [3], an examination of a pulverized coal flame in the same laboratory-scale boiler, albeit with a different burner, was conducted. Various RANS models, such as S-KE, R-KE, RNG-KE, SST, and RSM, were employed to simulate turbulent flow. The RSM model [25, 26] yielded the most accurate results when compared to experimental data. Additionally, for the 100-kW power burner, RSM exhibited superior performance in comparison to other RANS models. The standard wall function is applied to address near-wall turbulence. The P1 radiation model is employed to simulate the radiative heat transfer mechanism [27], and the Weighted Sum of Gray Gases Model [28] is utilized for calculating the absorption coefficient. Assumptions are made for particle emissivity (0.9) and wall emissivity (0.7). Upon entering a high-temperature environment, fuel particles undergo evaporation and pyrolysis. The volatile matter undergoes homogeneous combustion in the gas phase, while the residual char combusts through heterogeneous reactions.

The devolatilization model predicts the yield of volatile matter and employs a two competing rates model for biomass devolatilization. A kinetics/diffusion-limited model is adopted for char combustion, with the surface reaction rate being determined by either kinetics or diffusion rate [29, 30]. For biomass fuel, devolatilization and char combustion kinetic rates are sourced from Li et al. [31], based on similar ultimate and proximate analysis data. For coal, the kinetic rates are derived from Toporov et al. [32]. Gas-phase combustion is modelled using a global reaction scheme involving two irreversible reactions. Initially, volatile matter decomposes to CO and H_2O, followed by the oxidation of CO to CO_2 [33]. Finite rate/Eddy Dissipation is employed to simulate gas reactions in this study [34].

3 Experimental Setup and Fuel Properties

The experimental procedures were conducted using a test furnace at RWTH Aachen University [32]. Figure 1 illustrates the sectional view of swirl burner (100-kW) with all dimensional details.

The pulverized fuel is injected by using primary inlet (L2). A swirling flame is sent through secondary inlet (L4). The geometric swirl number is 0.96. There are also tertiary (L7) and staging air (L9) inlets. The experiments carried out under oxy-fuel environment. The measurement data of 100 kW coal combustion flame were provided from the experimental study [32]. The test facility is a vertical cylindrical furnace having 0.4 m diameter and 2.1 m length. The swirl burner is moveable up to down axially. The data are taken from four ports having a gate valve system located 2.1 m below from the top of boiler. The Rhenish lignite and torrefied biomass are used for the experimental study. The operating conditions are listed in Table 1.

Coal is introduced to the system at a rate of 6.5 kg/h. For an equivalent thermal power output, fuel blends should be supplied at calculated rates shown in Table 2.

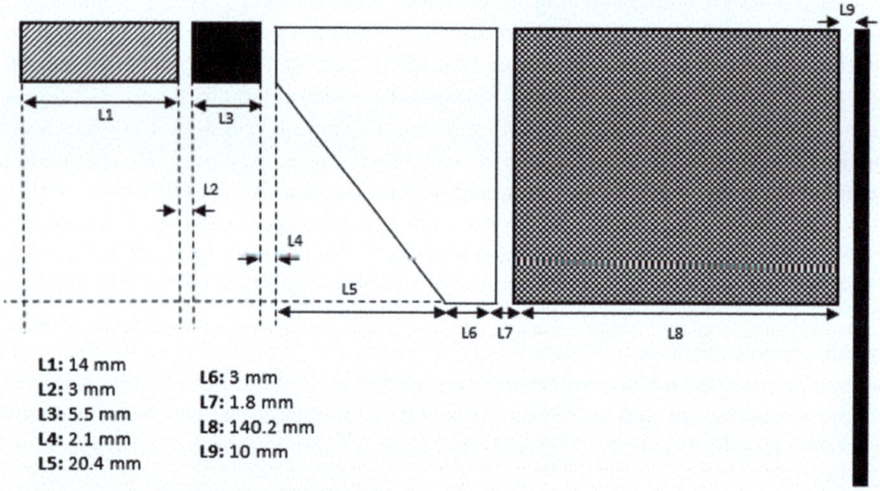

Fig. 1 The sectional view of swirl burner

Table 1 Operation conditions

	MRF (kg/h)	O_2 vol	CO_2 vol	T (K)
Coal	6.5	–	–	–
Primary inlet	17.6	0.19	0.81	313
Secondary inlet	26.6	0.21	0.79	333
Tertiary inlet	1.5	0.21	0.79	333
Staging inlet	54.9	0.21	0.79	1173

Table 2 Mass flow rate for selected three blending ratios

	25B–75C	50B– 50C	75B–25C
Mass Flow Rate (kg/h)	1.7–5.2	3.7–3.7	5.9–1.9

For blending ratio 25B–75C, 1.7 kg/h biomass and 5.2 kg/h coal should be supplied for co-combustion for similar power output while using 100% coal.

Table 3 presents the proximate and ultimate analysis of the fuel as received and dry-ash free. The lower and upper heating values are also shown.

Additionally, particle size distribution details can be found in Table 4. The measured particle size distribution, depicted in Fig. 2, is fitted with a Rosin–Rammler distribution. Q represents the mass fraction smaller than d.

It is evident that the alignment of Rosin–Rammler fits (CFD inputs) with experimental data is quite satisfactory. Nevertheless, there are some minor deviations, the impact of which may be disregarded in the results.

The devolatilization process for biomass is characterized by a two-competing rates model, as proposed by Kobayashi et al. [35], whereas a single rate model is employed for coal. For both fuels, a kinetics/diffusion-limited model is utilized to describe char combustion. Detailed kinetic data are provided in Table 5.

The problem is defined in a 2D-axisymmetric domain. The solution domain and boundary types are depicted in Fig. 3. Mass flow rate boundaries are designated for the primary, secondary, tertiary, and staging inlets. Inlets are assigned constant

Table 3 The details of fuel composition

	Coal		Biomass	
	AR	DAF	AR	DAF
Moisture [w–%]	8.4	–	2.26	–
Ash [w–%]	4.1	–	2.14	–
VM [w–%]	46.6	53.3	68.7	71.8
Char [w–%]	40.9	46.7	26.9	28.2
C [w–%]	67.4	77.0	55.2	57.7
H [w–%]	4.24	4.85	5.52	5.77
O [w–%]	14.7	16.8	34.6	36.2
N [w–%]	0.86	0.98	0.25	0.26
S [w–%]	0.30	0.34	0.03	0.03
LHV/HHV [MJ/kg]	25.1/26.3			19.1/20.4

Table 4 The particle size distribution of coal and biomass

	Coal	Biomass
Dv(10)	6 μm	5 μm
Dv(50)	25 μm	18 μm
Dv(90)	75 μm	45 μm

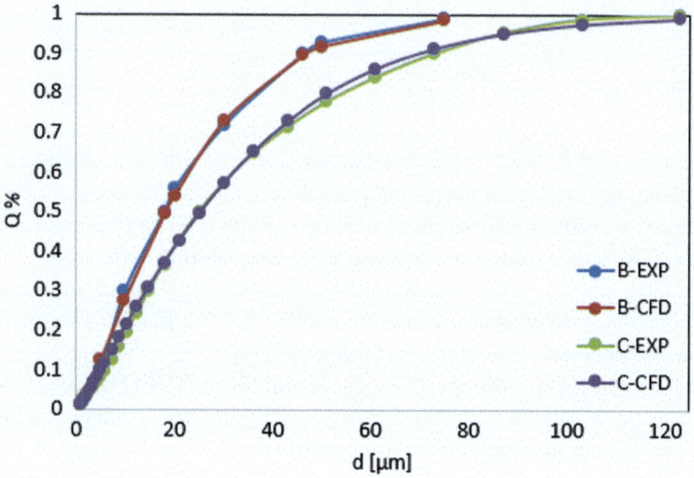

Fig. 2 The comparison of experimental input and CFD input of particles (C_EXP: Coal particle distribution in experimental study, C_CFD: Coal particle distribution input for CFD, B_EXP: Biomass particle distribution in experimental study, B_CFD: Biomass particle distribution input for CFD) [10]

Table 5 The kinetic data for coal and biomass

Devolatilization kinetics		
	Coal* (Rhenish lignite)	Biomass** (torrefied)
A [s^{-1}]	2×10^5	5.48×10^3
E [kJ/mol]	49	59.8
A [s^{-1}]	–	2.15×10^7
E [kJ/mol]	–	147
Char reaction kinetics		
	Coal (Rhenish lignite)	Biomass (torrefied)
A [s^{-1}]	2×10^{-5}	56.3
Ea [kJ/mol]	79.4	78

temperatures based on experimental data. A pressure outlet boundary condition is applied at the exit, and the remaining boundaries are treated as walls.

A mesh independence study was conducted in a prior investigation on the same furnace [3]. The results of this study, illustrating a generated grid consisting of 35,000 cells, are also presented in Fig. 2. Measurement data is collected at four different stations within the furnace, located at x = 0.025 m, 0.05 m, 0.2 m, and 0.3 m.

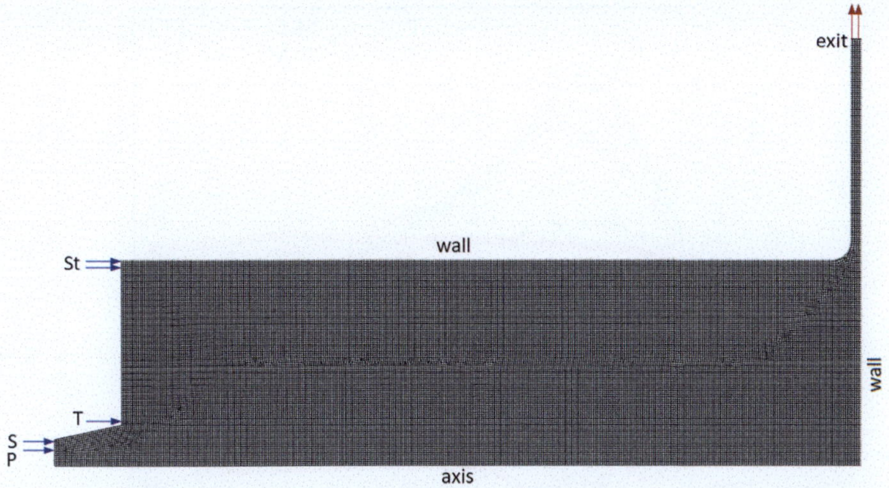

Fig. 3 The solution domain, BC types (P: Primary inlet, S: Secondary inlet, T: Tertiary inlet, St: Staging air) and generated grid

4 Results

Five different turbulence models have been employed to determine the model that best fits the flow measurements. In Fig. 4, the experimental data from x = 0.05 m, 0.2 m and 0.3 m are presented as a function of r. A detailed study was previously conducted [3], revealing that RSM exhibited superior performance in predicting turbulence compared to other TV models. It is evident that RSM captures the measured velocity values of the flame more accurately than other models. RSM is recognized for delivering reliable results in swirling flows in the literature [23–26]. It is seen that the RSM model exhibits the minimum error between experimental and calculated temperature data. The trends in temperature increase align well with the experimental data at the axial distance 0.05 m and 0.3 m.

Different co-firing ratios are being examined to identify the optimal blend for seamless integration into existing coal-fired boilers. Both coal and biomass are introduced through the same inlet for this investigation. Torrefied biomass is selected for its composition, which closely resembles coal in comparison to other types of biomass. The blending feeding rates are set at 25% Biomass + 75% Coal, 50% Biomass + 50% Coal, 75% Biomass + 25% Coal. The temperature fields for each blending ratio are illustrated in Fig. 5. Boiler geometry is taken until the axial distance of 0.5m which is the last measurement level in the experimental analysis.

It is evident from Fig. 5 that an increase in the percentage of torrefied biomass in the blend results in a reduction of peak flame temperatures. The specific values for peak flame temperature are 1557K for 25B–75%C, 1547K for 50B–50%C, 1522K for 75B–25%C. With the increased proportion of biomass in the blend, the flame

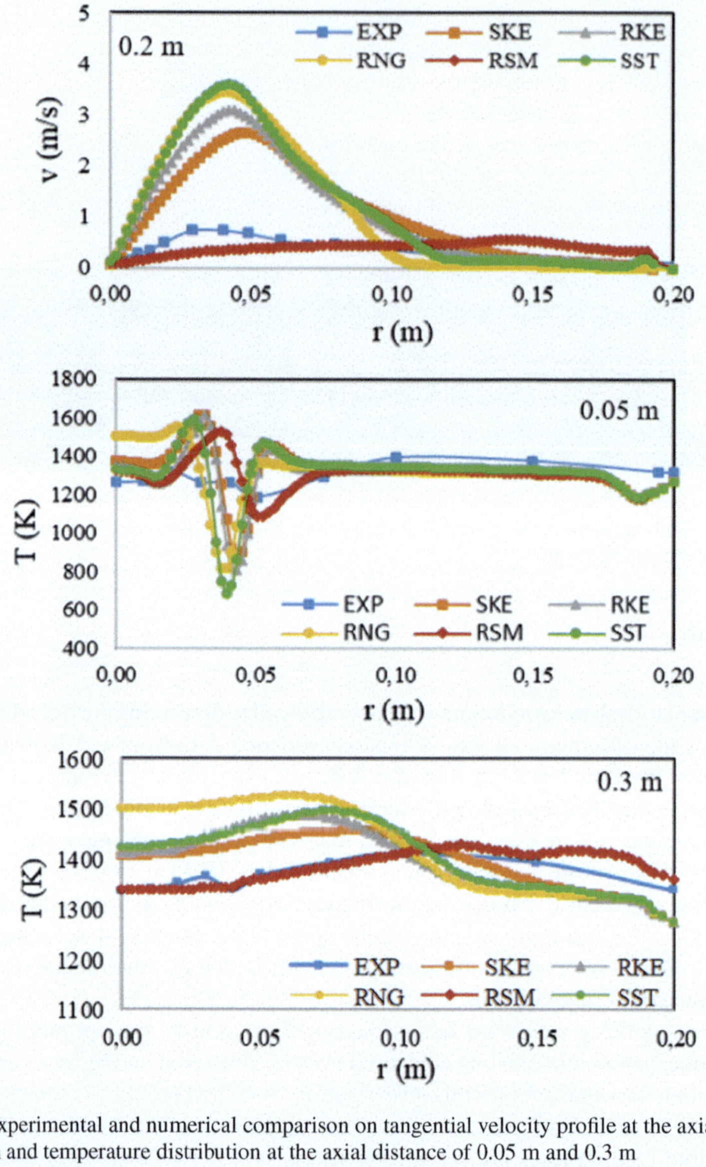

Fig. 4 Experimental and numerical comparison on tangential velocity profile at the axial distance of 0.02 m and temperature distribution at the axial distance of 0.05 m and 0.3 m

structure occurs later. As anticipated, the SO_2 (a) and CO_2 (b) mole fraction decreases with an increasing biomass content, as depicted in Fig. 6.

The discrete random walk model is employed for simulating turbulent particle dispersion, requiring a sufficient number of trials (M) to yield meaningful results. The computational cost is directly influenced by the number of trials, necessitating the identification of an optimal value to achieve reasonable results with minimal

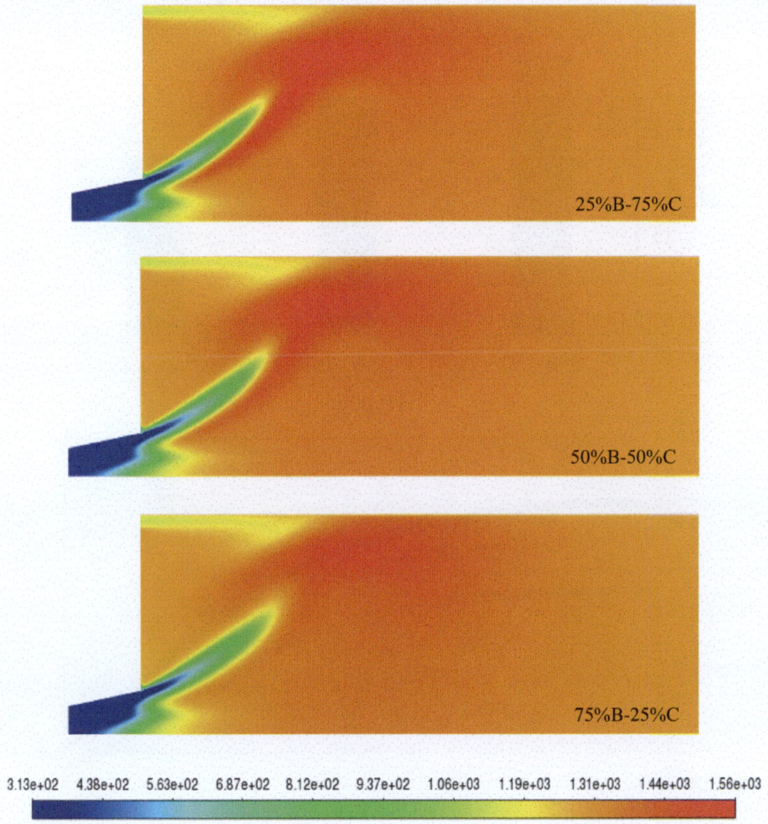

Fig. 5 Temperature profile (K) of three blending ratios (B: Biomass, C: Coal)

computational expense. In this study, different trial numbers (M) are tested to investigate their impact on the outcomes for co-firing combustion, with M values chosen as 1, 3, 5, 10, and 20. Temperature profiles along the axis are presented in Fig. 7.

It is evident from the results that M equal to 5 yields optimal outcomes when compared to other values. The temperature profile for M equals 20 is quite similar to $M = 10$. The number of particle sizes for coal and biomass is defined as 28 and 9, respectively.

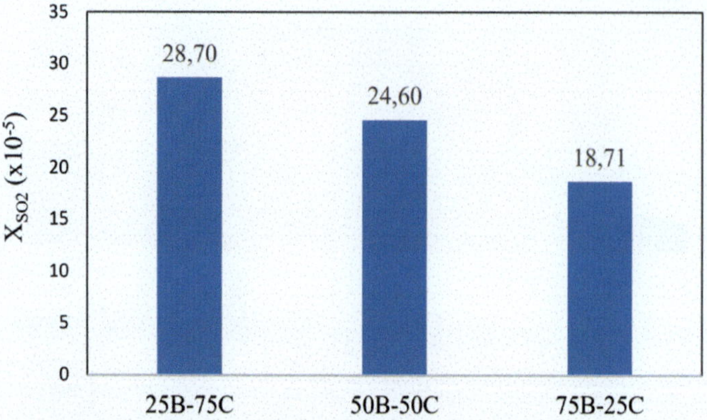

(a)

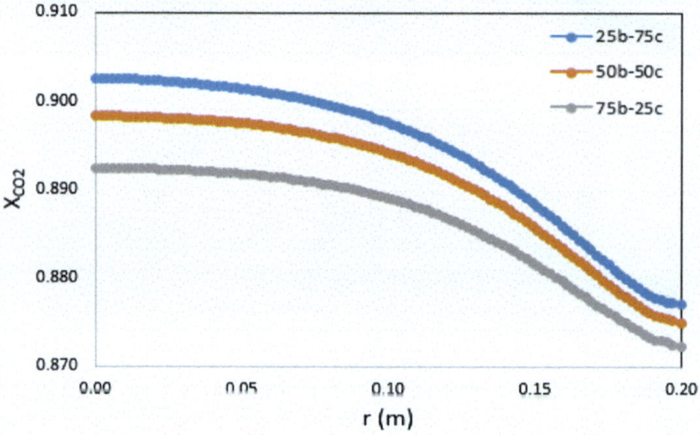

(b)

Fig. 6 SO_2 **a** and CO_2 **b** emissions

Fig. 7 Temperature profiles along the axis according to M number

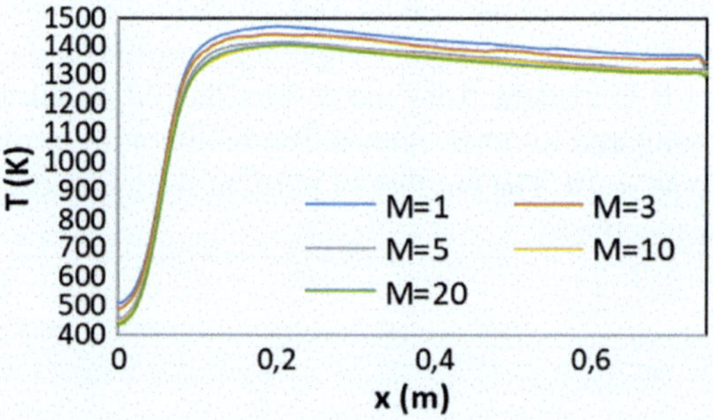

5 Conclusions

Numerical investigations were carried out on the pulverized coal and biomass flame using the Eulerian–Lagrangian model to simulate two-phase flow. Turbulence modelling was approached with five different RANS models: S-KE, RNG-KE, R-KE, SST, and RSM. The calculated axial velocity and static temperature values were then compared with experimental data. Results indicated that, overall, the RSM model exhibited better agreement with the experimental data. Additionally, the study delves into various co-firing ratios, revealing that an increase in the biomass ratio led to a reduction in the peak temperature of the flame, aligning with expectations for achieving lower emissions. To optimize computational cost while obtaining acceptable results, the number of trials in the random walk model of turbulent dispersion was set at $M = 5$.

References

1. Lackner, M., Winter, F., Agarwal, A.K. (eds.): Handbook of Combustion. Wiley, Chichester, UK (2010)
2. Benim, A.C., Deniz Canal, C., Boke, Y.E.: Computational investigation of oxy combustion of pulverized coal and biomass in a swirl burner. Energy. **238**(C), 121852 (2022)
3. Benim, A.C., Deniz Canal, C., Boke, Y.E.: A validation study for RANS based modelling of swirling pulverized fuel flames. Energies **14**(21), 7323 (2021)
4. Pfeiffelmann, B., Diederich, M., Gül, F., Benim, A.C., Heese, M., Hamberger, A.: Computational and experimental investigation of an industrial biomass furnace. Chem. Eng. Technol. **44**, 1538–1546 (2021)
5. Rosendahl, L. (ed.): Biomass combustion science, technology and engineering, Elsevier, Amsterdam, The Netherlands (2013)
6. Van L.S, Koppejan, J.: The Handbook of Biomass Combustion and Co-firing, Routledge, London, UK (2012)
7. Fatehi, H., Bai, X.S: Gasification and combustion of biomass: physical description and mathematical modelling. In: Handbook of Clean Energy Systems, Wiley, New York, USA (2015)
8. Benim, A.C., Zinser, W.: A segregated formulation of Navier-stokes equations with finite elements. Comput. Methods Appl. Mech. Eng. **57**(2), 223–237 (1986)
9. Ferziger, J.H., Peric, M.: Computational Methods for Fluid Dynamics, 3rd edn. Springer, Berlin, Germany (2002)
10. ANSYS Fluent Theory Guide. Rel. 2018, ANSYS Inc., Canonsburg, USA (2018)
11. Maliska, C.R.: Fundamentals of Computational Fluid Dynamics. Springer, Berlin, Germany (2023)
12. Versteeg, H.K, Malalasekera, W.: An Introduction to Computational Fluid Dynamics, 2nd ed. Pearson, Harlow, England (2007)
13. Moukalled, F., Mangani, L., Darwish, M.: The Finite Volume Method in Computational Fluid Dynamics. Springer, Berlin, Germany (2016)
14. Crowe, C.T., Schwarzkopf, J.D., Sommerfeld, M., Tsuji, Y.: Multiphase Flows with Droplets and Particles, 2nd edn. CRC Press, Boca Raton, USA (2012)
15. Brennen, C.E.: Fundamentals of Multiphase Flow. Cambridge University Press, Cambridge, UK (2009)
16. Schlichting, H.: Boundary Layer Theory, 6th edn. McGraw-Hill, New York, USA (1968)

17. Bird, R.B., Stewart, W.E., Lightfoot, E.N.: Transport Phenomena. Wiley, New York (1960)
18. Gosman, A.D., Ioannides, E.: Aspects of computer simulation of liquid fuelled combustors. J. Energy **7**, 482–490 (1983)
19. Lefebvre, A.H., Ballal, D.R.: Gas Turbine Combustion, 3rd edn. Taylor & Francis, Boca Raton, USA (2010)
20. Lefebvre, A.H., McDonnel, V.G.: Atomization and Sprays, 2nd edn. CRC Press, Boca Raton, USA (2017)
21. Durbin, P.A., Pettersson Reif, B.A.: Statistical Theory and Modeling for Turbulent Flows, 2nd edn. Wiley, New York, USA (2011)
22. Leschziner, M.: Statistical Turbulence Modelling for Fluid Dynamics—Demystified. Imperial College Press, London, UK (2016)
23. Benim, A.C.: Finite element analysis of confined turbulent swirling flows. Int. J. Numer. Meth. Fluids **11**(6), 697–717 (1990)
24. Xia, J.L., Smith, B.L., Benim, A.C., Schmidli, J., Yadigaroglu, G.: Effect of inlet and outlet boundary conditions on swirling flows. Comput. Fluids **26**(8), 811–823 (1997)
25. Launder, B.E., Reece, G.J., Rodi, W.: Progress in the development of a reynolds-stress turbulent closure. J. Fluid Mech. **68**(3), 537–566 (1975)
26. Speziale, C.G., Sarkar, S., Gatski, T.B.: Modeling the pressure-strain correlation of turbulence: an invariant dynamical systems approach. J. Fluid Mech. **227**, 245–272 (1991)
27. Özisik, N.: Radiative Transfer. Wiley, New York, USA (1973)
28. Smith, T.F., Shen, Z.F., Friedman, J.N.: Evaluation of coefficients for the weighted sum of gray gases model. J. Heat Transfer **104**, 602–608 (1982)
29. Badzioch, S., Hawskley, P.G.W.: Kinetics of thermal decomposition of pulverized coal particles. Int. Eng. Chem. Proc. Des. Develop. **9**, 521–530 (1970)
30. Baum, M.M., Street, P.J.: Predicting the combustion behaviour of coal particles. Combust. Sci. Technol. **3**, 231–243 (1971)
31. Li, J., Brzdekiewicz, A., Yang, W., Blasiak, W.: Co-firing based on biomass torrefaction in a pulverized coal boiler with aim of 100% fuel switching. Appl. Energy **99**, 344–354 (2012)
32. Toporov, D., Bocian, P., Heil, P., Kellermann, A., Stadler, H., Tschunko, S., Förster, M., Kneer, R.: Detailed investigation of a pulverized fuel swirl flame in CO_2/O_2 atmosphere. Combust. Flame **155**, 605–618 (2008)
33. Turns, S.R.: An Introduction to Combustion. McGraw-Hill, New York, USA (2012)
34. Magnussen, B.F., Hjertager, B.H.: On mathematical modelling of turbulent combustion on special emphasis on soot formation and combustion. Symp (Int) Combust **16**,719–729 (1976)
35. Kobayashi, H., Horward, J.B., Sarofirm, A.F.: Coal devolatilization at high temperatures. Symp (Int) Combust **16**(1), 411–425 (1977)

Heat Transfer in Tangent Hyperbolic Nanofluid Flow Over a Stretching Sheet with Convective Boundary

Ankita Bisht and Rajesh Sharma

Abstract In this study, boundary layer flow of a Tangent hyperbolic nanofluid (THNF) over a stretching sheet in presence of magnetic field is numerically investigated. The fluid is considered incompressible, with thermophoresis and Brownian diffusion effects included. Dimensional governing equations for fluid are transformed into dimensionless ordinary differential equations using appropriate similarity transformations. These equations are then solved using the finite difference method via MATLAB's bvp4c routine. The study examines the impact of various flow parameters on velocity, temperature, nanoparticle concentration, illustrated through graphs. Furthermore, local skin friction and Nusselt number are computed and analysed graphically. It is found that both the Weissenberg (We^*) and Hartmann (M) number reduce fluid motion. Moreover, increase in the heat transfer rate results from an increase in the convective Biot number value whereas an opposite relation is observed with Hartmann number (M) and Weissenberg number (We^*).

Keywords Tangent hyperbolic fluid · Magnetohydrodynamic · Convective boundary condition · Finite difference method

1 Introduction

The study of non-Newtonian fluids has gained significant attention over the past two decades, becoming a crucial area of research. This heightened interest stems from the need to understand the thermo-physical characteristics of these fluids, driven by their

A. Bisht (✉)
Department of Mathematics, Amity School of Physical Sciences, Amity University Punjab, Mohali, India
e-mail: bishtankita2@gmail.com

R. Sharma
Department of Mathematics, North Eastern Hill University, Shillong, Meghalaya, India
e-mail: raj.juit@gmail.com

R. K. Sharma et al. (eds.), *Mathematical Theory and Simulation of Scientific Problems*, Springer Proceedings in Mathematics & Statistics 487,
https://doi.org/10.1007/978-981-96-2579-6_4

wide-ranging industrial and technological applications. As a result, there is an extensive body of literature on analytical and numerical solutions related to non-Newtonian fluids. However, investigating non-Newtonian fluids presents several serious challenges. The governing equations for non-Newtonian fluids are complex and nonlinear, with no universal constitutive equation to describe their varied behavior. To address this, multiple models have been developed, among which the tangent hyperbolic fluid (THF) model is particularly important. Laboratory experiments have shown that this model accurately predicts the shear-thinning behavior, a common characteristic of many non-Newtonian fluids, making it particularly useful for practical applications.

Unlike Newtonian fluids, whose viscosity remains constant regardless of the applied shear rate, the viscosity of THFs decreases with increasing shear rate. This model is particularly useful for accurately predicting the flow behavior of fluids in industrial applications where precise control over viscosity is essential. As a result, many researchers have employed the constitutive equations of the tangent hyperbolic model to tackle a range of physical problems. In their study, Akbar et al. [1] investigated the MHD boundary layer flow of a THF over a Stretching Sheet and found that Weissenberg number enhances the thickness of the fluid. Malik et al. [2] investigated the MHD flow of a THF around a stretched cylinder using the Keller box method. Akbar et al. [3] analysed the steady blood flow through a tapered artery with a radially symmetric, axially nonsymmetric mild stenosis, considering the effects of heat and mass transfer and treating blood as a THF. Waqas et al. [4] studied the behavior of THF on a nonlinear stretched sheet with variable thickness using non-Fourier flux theory for energy expression. They found that velocity decreases with larger material power law indices and Weissenberg numbers.

Tangent hyperbolic nanofluid (THNF) refers to a blend of nanosized particles with a non-Newtonian THF. Adding nanoparticles enhances thermal conductivity, leading to improved thermal processes. This emerging topic has attracted significant attention from researchers due to its wide-ranging applications and potential benefits. The THNF is widely used in the cooling of electronic components, which generate significant heat during operation. Its enhanced thermal conductivity allows for more efficient heat dissipation, thereby reducing the risk of overheating and potential component failure. Choi and Eastman [5] were the pioneers in proposing that the addition of nanoparticles (1–100 nm) to conventional base fluids can significantly enhance their thermal conductivity. Amer et al. [6] examined the flow characteristics and properties of THNF over a stretching sheet embedded in a porous medium. Khan et al. [7] explored the 2D flow of an MHD THF containing nanoparticles over a stretching surface.

Flow over a stretching sheet refers to a fluid dynamics scenario where a fluid flows over a surface that is being stretched in one or more directions. This type of flow is commonly studied in the context of boundary layer theory, where the stretching of the sheet can induce complex flow behavior in the fluid, especially when combined with factors like heat transfer, magnetic fields, or non-Newtonian fluid properties. Understanding this flow is crucial in many industrial and engineering applications, such as in the manufacturing of materials, cooling of electronic devices, and the processing of polymer sheets. Sakiadis [8] was the pioneer in this field, investigating boundary layer flow with uniform speed over a continuous solid surface. Crane [9] expanded on this concept by describing viscous fluid flow resulting from the linear stretching of a smooth sheet, notably achieving a similar solution. Cortell [10] explored the flow of non-Newtonian fluids over stretching surfaces and obtained numerical solutions to the governing equations using the shooting method. Recently, the study of stretched flows of tangent hyperbolic fluids (THFs) has gained attention from many researchers [11–14].

A review of THF flow literature highlights that boundary layer flow of THNF under a magnetic field over a linearly stretching sheet with convective boundary conditions remains underexplored. This gap prompted the author to investigate magnetohydrodynamic THF with nanoparticles on such a sheet. The study employs MATLAB's bvp-4c tool to solve the governing flow equations across various parameters. Key physical quantities are computed, analyzed, and presented in both graphical and tabular forms under different conditions.

2 Mathematical Modelling

We discuss a 2D, steady, viscous and incompressible flow of THNF over a stretching sheet at $y = 0$ with velocity $\tilde{U}_s(x) = \tilde{a}\, x$ and convective boundary condition. The flow is confined in $y > 0$ and B_0 strength magnetic field is used in a direction perpendicular to the flow. The schematic representation of the present fluid flow problem is given via Fig. 1. Based on the foregoing assumptions, the governing equations are as follows (see Refs. [1, 7]).

$$\frac{\partial \tilde{u}}{\partial x} + \frac{\partial \tilde{v}}{\partial y} = 0 \tag{1}$$

$$\tilde{u}\frac{\partial \tilde{u}}{\partial x} + \tilde{v}\frac{\partial \tilde{u}}{\partial y} = \nu\left[(1-\gamma) + \gamma\lambda\sqrt{2}\left(\frac{\partial \tilde{u}}{\partial y}\right)\right]\frac{\partial^2 \tilde{u}}{\partial y^2} - \frac{\sigma B_0^2 \tilde{u}}{\rho} \tag{2}$$

$$\tilde{u}\frac{\partial \tilde{T}}{\partial x} + \tilde{v}\frac{\partial \tilde{T}}{\partial y} = \kappa\frac{\partial^2 T}{\partial y^2} + \tau\left(D_B\left(\frac{\partial \tilde{\phi}}{\partial y}\frac{\partial \tilde{T}}{\partial y}\right) + \frac{D_T}{T_\infty}\left(\frac{\partial \tilde{T}}{\partial y}\right)^2\right) \tag{3}$$

$$\tilde{u}\frac{\partial \tilde{C}}{\partial x} + \tilde{v}\frac{\partial \tilde{C}}{\partial y} = D_B\frac{\partial^2 \tilde{C}}{\partial y^2} + \frac{D_T}{T_\infty}\left(\frac{\partial^2 \tilde{T}}{\partial y^2}\right) \tag{4}$$

The relevant boundary conditions are:

$$\left.\begin{array}{c} \tilde{u} = \tilde{U}_s(x) = \tilde{a}x,\ \tilde{\kappa}\frac{\partial \tilde{T}}{\partial y} = -\tilde{h}_w(\tilde{T}_w - \tilde{T}),\ \tilde{C} = \tilde{C}_w at\, y = 0 \\ \tilde{u} \to 0,\ \tilde{T} \to \tilde{T}_\infty,\ \tilde{C} = \tilde{C}_\infty\, as\, y \to \infty \end{array}\right\} \tag{5}$$

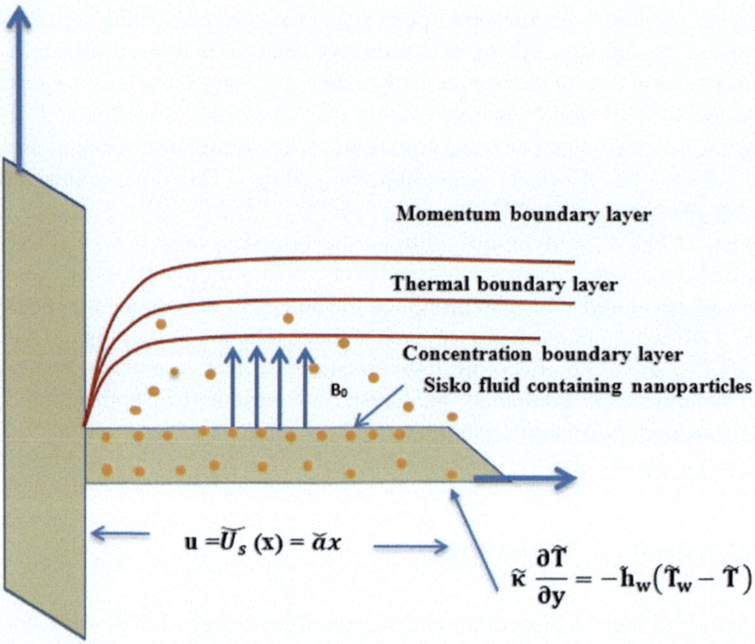

Fig. 1 Physical representation of the flow problem

The aforementioned equations include the following: $\tilde{C}$ is nanoparticle concentration; $\tilde{T}_w$ is surface temperature and $\tilde{C}_w$ stands for surface concentration; ρ is the base fluid density; $\tilde{T}_\infty$ signifies ambient temperature and $\tilde{C}_\infty$ ambient concentration; γ is the power law index; α is thermal diffusivity, λ is time constant. Additionally, D_B is Brownian diffusion coefficient, while D_T is the thermophoresis constant; ν represent kinematic viscosity; $\tilde{h}_w$ denotes heat transfer coefficient; and κ is the thermal conductivity.

Similarity transformations:

$$\xi = \sqrt{\frac{\tilde{a}}{\mathrm{v}}}\mathrm{y},\ \psi = \sqrt{(\tilde{a}\mathrm{vx})}F(\xi),\ \tilde{u} = \tilde{a}\mathrm{x},\ \Theta(\xi) = \frac{\tilde{T} - \tilde{T}_\infty}{\tilde{T}_w - \tilde{T}_\infty},\ \Phi(\xi) = \frac{\tilde{C} - \tilde{C}_\infty}{\tilde{C}_w - \tilde{C}_\infty} \quad (6)$$

On applying the similarity transformations to Eqs. (1–5), they are transformed into the following form:

$$\left((1-\gamma) + \gamma We^* F''\right)F''' - F'^2 - M^{*^2} F' = 0 \quad (7)$$

$$\Theta'' + \overset{*}{\mathrm{Pr}}(\mathrm{Nb}^*\Theta'\Phi' + F\Theta' + \mathrm{Nt}^*\left(\Theta'\right)^2 = 0 \quad (8)$$

$$\Phi'' + \overset{*}{\Pr}\,\mathrm{Le}^* F\Phi' + \frac{\mathrm{Nt}^*}{\mathrm{Nb}^*}\Theta'' = 0 \tag{9}$$

With

$$\left.\begin{array}{l} F'(0) = 1, F(0) = 0,\ \Theta'(0) = 1,\ \theta'(0) \\ \quad + \beta(1 - \theta(0)) = 0, \Phi(0) = 1 \\ \quad F'(\xi) \to 0,\ \Phi(\xi) \to 0, \Theta(\xi) \to 0, as\,\xi \to \infty \end{array}\right\} \tag{10}$$

Here, the symbols We^*, M^*, $\overset{*}{\Pr}$, Nt^*, Nb^*, Le^* stands for Weissenberg number, Hartman number, Prandtl number, Thermophoresis, Brownian diffusion, and Lewis number, respectively and their definition are:

$$\begin{array}{l} We^* = \lambda x\sqrt{\dfrac{2\tilde{a}}{\nu^3}}, M^* = \dfrac{\sigma B_0^2}{\rho\tilde{a}}, \Pr = \dfrac{\nu}{\alpha}, Nb^* = \dfrac{\tau D_B(\tilde{C}_w - \tilde{C}_\infty)}{\alpha}, \\ Le^* = \dfrac{\alpha}{D_B}, Nt^* = \dfrac{\tau D_T(\tilde{T}_w - \tilde{T}_\infty)}{\alpha\tilde{T}_\infty}, \end{array} \tag{11}$$

The quantities of physical importance i.e. local Nusselt number (LNN) and Local Skin friction coefficient (LSFC) at the surface in non-dimensional form is given as:

$$\mathrm{Nu}/\sqrt{\mathrm{Re_x}} = -\Theta'(0) \tag{12}$$

$$\sqrt{\mathrm{Re_x}}C_f = \left((1-\gamma)F'' + \frac{\gamma}{2}\mathrm{We}F''^2\right)|_{\xi=0} \tag{13}$$

where $\mathrm{Re_x} = \frac{\tilde{U}_s x}{\nu}$.

3 Result and Discussion

The numerical solutions for Eqs. (7–9), in conjunction with Eq. (10), were obtained using the bvp-4c solver in MATLAB. The bvp-4c tool, known for its robustness in solving boundary value problems, was employed with a step size of $\Delta\xi = 0.05$, ensuring precise approximation of the solution across the domain. To account for the behavior of the system at infinity, the boundary condition $\xi \to \infty$ was approximated with a maximum value of $\xi_{max} = 10$, which provided a sufficient range for accurate analysis. An error tolerance of 10^{-6} was meticulously maintained for all simulations, ensuring the reliability and accuracy of the numerical results.

To facilitate a deeper understanding of the underlying physical phenomena, the outcomes of the numerical simulation are comprehensively illustrated through a series of graphs. These graphical representations offer clear insights into the behavior of the tangent hyperbolic nanofluid (THNF) under various conditions, enabling a thorough examination of the effects of key parameters on velocity, temperature, and concentration profiles.

Figures 2, 3, and 4 show how the Weissenberg number (We^*) affects the velocity, concentration, and temperature profiles of THNF. An increase in We^* leads to a decrease in the velocity profile, consistent with the increased relaxation time and higher resistance to flow. Conversely, both the temperature and concentration profiles increase with rising We^*, as the enhanced relaxation time allows for greater heat retention and nanoparticle accumulation. These graphs effectively illustrate the influence of We^*, on the fluid's flow and thermal behaviour.

Figures 5, 6, and 7 show the effect of the power law index (γ) on the velocity, temperature, and concentration profiles of the tangent hyperbolic nanofluid (THNF).

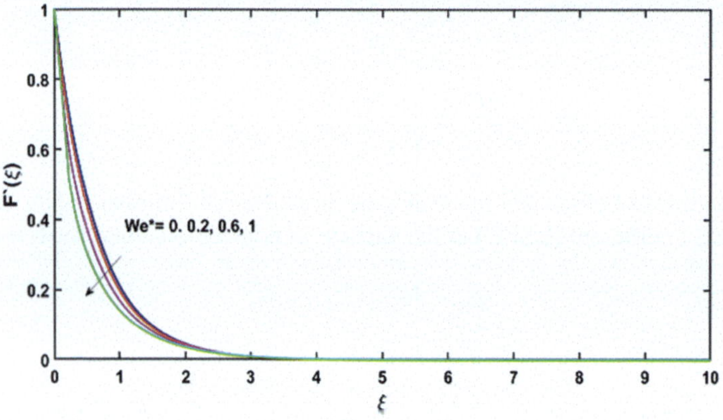

Fig.2 Velocity profile with We*

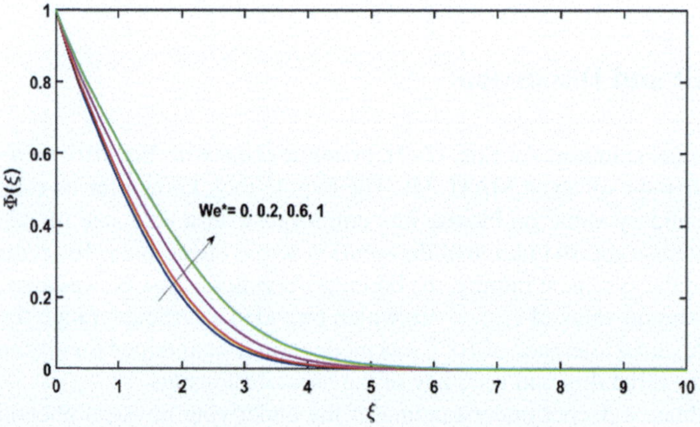

Fig. 3 Concentration profile with *We**

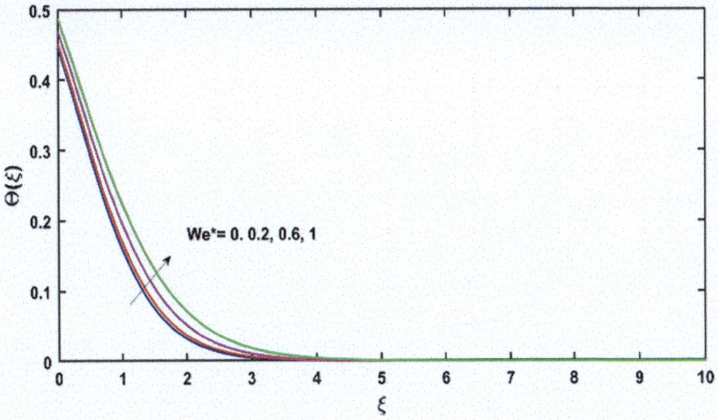

Fig. 4 Temperature profile with We*

As γ increases, both the velocity and the boundary layer thickness decrease. This happens because a higher γ indicates a shift from shear-thinning to shear-thickening behavior, which increases fluid resistance. Consequently, the temperature and concentration profiles increase as γ rises, due to reduced heat and nanoparticle dispersion in the more resistant fluid.

Figure 8 demonstrates the effect of the Biot number (β) on the temperature profile. The Biot number represents the ratio of convective heat transfer at the surface to conductive heat transfer within the surface of a body. As β increases, indicating stronger convective heat transfer relative to conduction, the temperature profiles also rise. This suggests that enhanced convection improves heat dissipation, leading to higher temperatures in the fluid near the surface.

Figures 9 and 10 illustrate the behavior of the local skin friction coefficient (LSFC) for varying values of the Weissenberg number (We^*) and the power law index (γ).

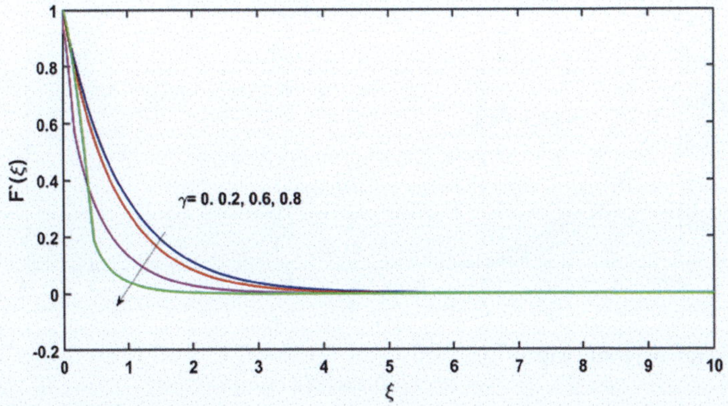

Fig. 5 Velocity profile with γ

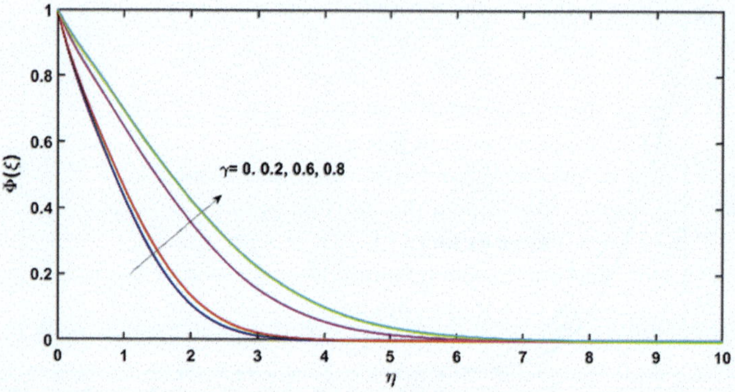

Fig. 6 Concentration profile with ϒ

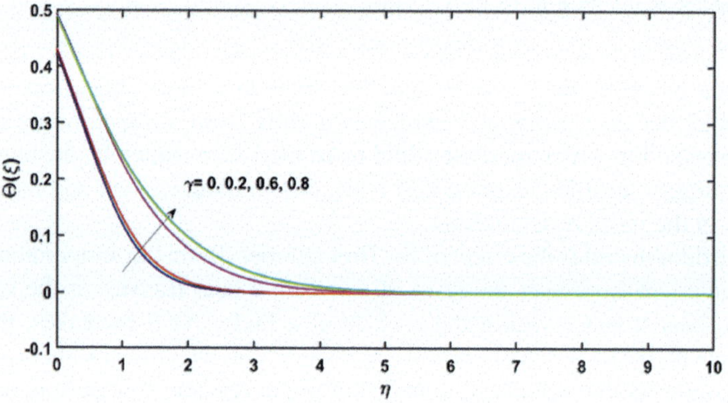

Fig. 7 Temperature profile with ϒ

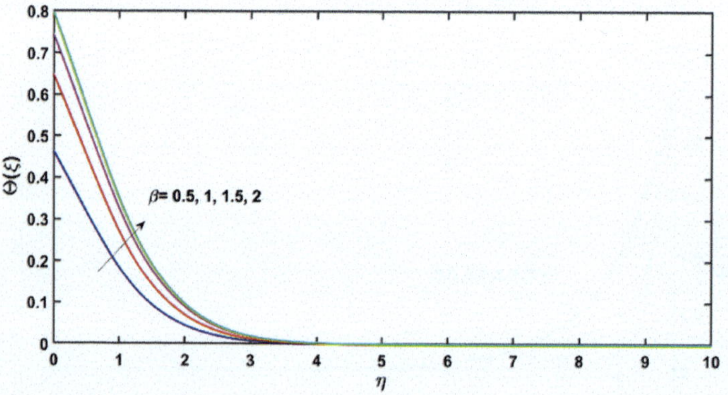

Fig. 8 Temperature profile with β

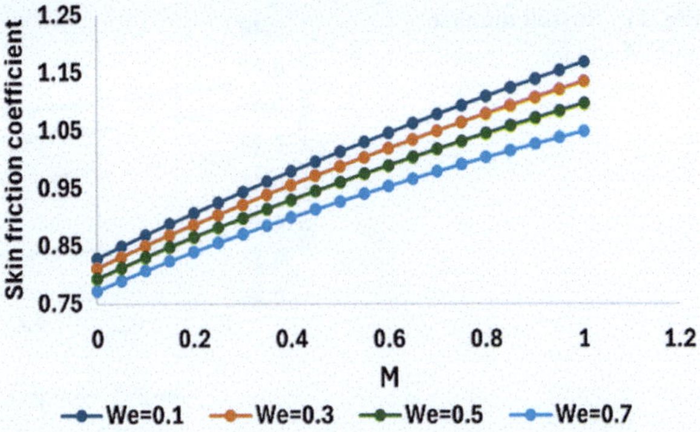

Fig. 9 Skin friction with M and *We**

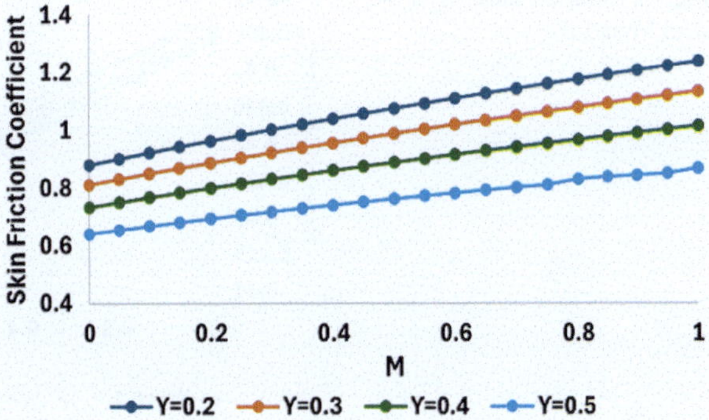

Fig. 10 Skin friction with M and ϒ

The data reveals that as the *We** and γ values increase, the absolute values of LSFC decrease. This indicates that higher *We** and γ reduce the fluid's resistance to shear, thereby lowering the skin friction at the surface. Additionally, the LSFC shows contrasting behavior with respect to the Hartmann number (M); as M increases, the LSFC behaves differently compared to its response to changes in *We** and γ.

Figures 11 and 12 depict the variation of the LNN with different values of the M and *We**, and the convective Biot number (β). The results indicate that an increase in the convective Biot number (β) leads to a rise in LNN, reflecting improved heat transfer efficiency with stronger convection. Conversely, LNN shows a decreasing trend with increasing Hartmann number (M) and Weissenberg number (*We**), suggesting that higher magnetic fields and fluid relaxation effects hinder heat transfer efficiency. These figures provide a comprehensive view of how these parameters affect the heat transfer characteristics of the THNF.

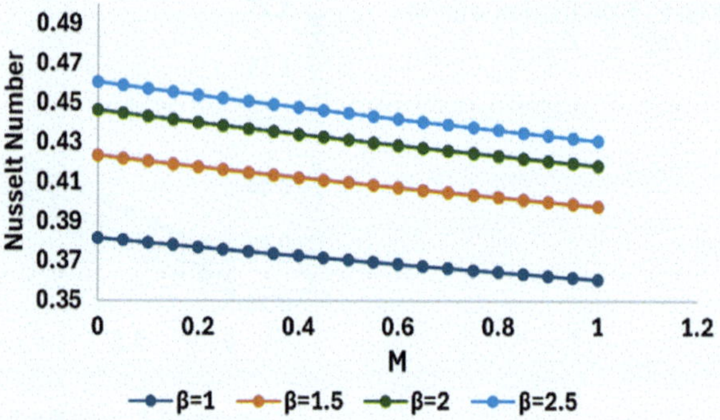

Fig. 11 Nusselt number with M and β

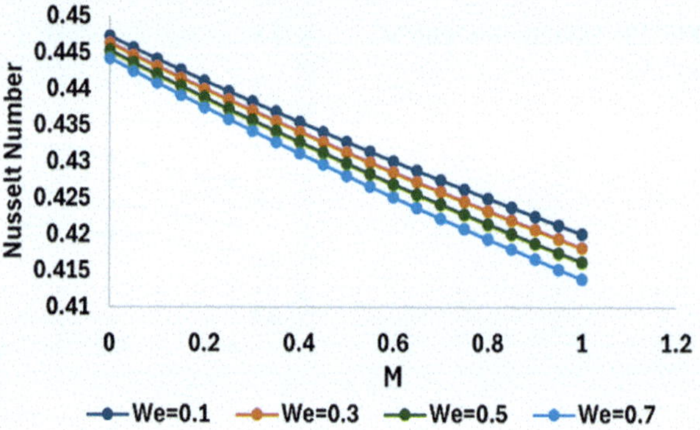

Fig. 12 Nusselt number with M and *We**

4 Conclusion

This study has presented a comprehensive analysis of the MHD boundary layer flow of a tangent hyperbolic nanofluid (THNF) over a stretching sheet with convective boundary conditions, incorporating the significant effects of thermophoresis and Brownian diffusion. The findings of this research contribute to a deeper understanding of how various parameters influence the behavior of THNFs in the presence of a magnetic field.

Key conclusions drawn from this investigation include:

1. **Velocity Profile**: The velocity of the THNF is found to decrease with increasing values of the Weissenberg number (We^*) and the power law index (γ). This indicates that higher elastic effects and non-Newtonian fluid characteristics tend to resist fluid motion. In contrast, the temperature and concentration profiles increase under the same conditions, suggesting that the fluid retains more thermal energy and nanoparticles, possibly due to reduced convective cooling.
2. **Temperature Distribution**: An increase in the convective parameter (β) leads to a significant rise in the temperature of the THNF. This behavior highlights the role of convective heat transfer at the boundary, where higher β values enhance the

thermal boundary layer, resulting in a more pronounced temperature distribution within the fluid.

3. **Skin Friction Coefficient**: The local skin friction coefficient (LSFC) shows an increasing trend with the Hartmann number (M), reflecting the influence of a stronger magnetic field in suppressing fluid motion and enhancing friction at the boundary. Conversely, the LSFC decreases with higher We^* and γ, emphasizing the damping effects of these parameters on frictional forces.
4. **Nusselt Number**: The local Nusselt number (LNN), a measure of the heat transfer rate, decreases with increasing values of M and We^*, indicating that stronger thermal stratification and elastic effects reduce the overall heat transfer efficiency. However, an increase in the convective parameter (β) leads to a rise in the Nusselt number, suggesting that convective heat transfer at the boundary is more effective under these conditions.

These findings not only deepen the understanding of THNF behaviors under the influence of magnetic fields but also provide a foundation for optimizing industrial processes involving nanofluids. The results have practical implications for applications where precise control of heat transfer and fluid flow is critical, such as in cooling systems, material processing, and energy devices. Future research could build on this work by exploring more complex geometries, time-dependent flows, or the inclusion of additional physical effects such as radiation or chemical reactions.

References

1. Akbar, N.S., Nadeem, S., Haq, R.U., Khan, Z.H.: Numerical solutions of magnetohydrodynamic boundary layer flow of tangent hyperbolic fluid towards a stretching sheet. Indian J. Phys. **87**, 1121–1124 (2013)
2. Malik, M.Y., Salahuddin, T., Hussain, A., Bilal, S.: MHD flow of tangent hyperbolic fluid over a stretching cylinder: using Keller box method. J. Magn. Magn. Mater. **395**, 271–276 (2015)
3. Akbar, N.S., Nadeem, S., Ali, M.: Influence of heat and chemical reactions on hyperbolic tangent fluid model for blood flow through a tapered artery with a stenosis. Heat Transf. Res. **43**(1) (2012)
4. Waqas, M., Bashir, G., Hayat, T., Alsaedi, A.: On non-fourier flux in nonlinear stretching flow of hyperbolic tangent material. Neural Comput. Appl. **31**, 597–605 (2019)
5. Choi, S.U., Eastman, J.A.: Enhancing thermal conductivity of fluids with nanoparticles (No. ANL/MSD/CP-84938; CONF-951135–29). Argonne National Lab. (ANL), Argonne, IL (United States) (1995)
6. Amer, A.M., Al Rashdi, S.A., Ghoneim, N.I., Megahed, A.M.: Tangent hyperbolic nanofluid flowing over a stretching sheet through a porous medium with the inclusion of magnetohydrodynamic and slip impact. Results Eng. **19**, 101370 (2023)
7. Khan, M., Hussain, A., Malik, M.Y., Salahuddin, T., Khan, F.: Boundary layer flow of MHD tangent hyperbolic nanofluid over a stretching sheet: a numerical investigation. Results Phys. **7**, 2837–2844 (2017)
8. Sakiadis, B.C.: Boundary-layer behavior on continuous solid surfaces. AIChE J. **7**, 26–28 (1961)
9. Crane, L.J.: Flow past a stretching plate. Z. Für Angew. Math. Und Phys. (ZAMP) **21**, 645–647 (1970)

10. Cortell, R.: Viscous flow and heat transfer over a nonlinearly stretching sheet. Appl. Math. Comput. **184**(2), 864–873 (2007)
11. Ibrahim, W.: Magnetohydrodynamics (MHD) flow of a tangent hyperbolic fluid with nanoparticles past a stretching sheet with second order slip and convective boundary condition. Results Phys. **7**, 3723–3731 (2017)
12. Alkaoud, A., Khader, M.M., Eid, A., Megahed, A.M.: Numerical simulation of the flow of a tangent hyperbolic fluid over a stretching sheet within a porous medium, accounting for slip conditions. Heliyon **10**(8) (2024)
13. Usman, M., Zubair, T., Hamid, M., Haq, R.U., Khan, Z.H.: Unsteady flow and heat transfer of tangent-hyperbolic fluid: legendre wavelet-based analysis. Heat Transf. **50**(4), 3079–3093 (2021)
14. Zeb, S., Khan, S., Ullah, Z., Yousaf, M., Khan, I., Alshammari, N., Hamadneh, N.N.: Lie group analysis of double diffusive MHD tangent hyperbolic fluid flow over a stretching sheet. Math. Probl. Eng.Probl. Eng. **2022**(1), 9919073 (2022)

Results of Asymptotic Analysis of an Elliptic Equation

Attou A. Miloua

Abstract Suppose $\Omega \subset \mathbb{R}^2$ a region, and consider the general elliptic equation

$$\Delta w = g\,(w)\,,$$

where g is a positive continuous function satisfying

$$\lim_{w \to 0^+} g\,(w) = \infty.$$

In the context of thin film equations, a solution w is classified as a point rupture solution, if there exists a point $p \in \Omega$, such that $w\,(p) = 0$ and $w\,(p) > 0$ in $\Omega \backslash \{p\}$. We aim to analyze the asymptotic behavior of radial solutions. Specifically, our primary objective is to investigate how the limiting profile of radial solutions w and their corresponding energies vary as a function of the prescribed volume.

Keywords Thin film · Point rupture solution · Radial solution · Singular elliptic equation · Quasi-linear elliptic equation · Asymptotic analysis

1 Foundation

This article focuses on radial solutions of a general elliptic equation

$$\Delta w = g\,(w) \tag{1}$$

within an open connected set $\Omega \subset \mathbb{R}^2$ subject to Neumann boundary conditions $\frac{\partial w}{\partial n} = 0$ on $\partial\Omega$. The controlling function g is assumed to be continuous non-negative on $(0, \infty)$ with the property

A. A. Miloua (✉)
Department of Chemistry, Mathematics and Physics, Pennsylvania Western University, California, PA 15419, USA
e-mail: miloua@pennwest.edu

R. K. Sharma et al. (eds.), *Mathematical Theory and Simulation of Scientific Problems*, Springer Proceedings in Mathematics & Statistics 487,
https://doi.org/10.1007/978-981-96-2579-6_5

$$\lim_{w \to 0^+} g(w) = \infty. \tag{2}$$

Motivated by the dynamics of thin film equations, we define a point rupture solution of (1) as a solution w, where $w(p) = 0$ for some $p \in \Omega$, and $w(x) > 0$ for any $x \in \Omega \setminus \{p\}$. Our primary objective is to perform an asymptotic analysis of these radial rupture solutions and their associated energies. In the context of fluid dynamics, particularly for lubrication models of thin films, w represents the thickness of a thin fluid layer over a planar region. The evolution of such films can often be described by the fourth-order partial differential equation:

$$w_t = -\nabla * (w^m \nabla w) - \nabla * (w^n \nabla \Delta w). \tag{3}$$

This class of equations has been extensively studied, with different parameter choices corresponding to various physical phenomena. For instance:

When n = m = 1, the equation models a thin jet in a Hele-Shaw cell [1, 4, 6, 7, 14].

When n = m = 3, it describes fluid droplets suspended from a ceiling [8].

When n = 0 and m = 1, it represents a modified Kuramoto–Sivashinsky equation relevant to the solidification of hypercooled melts [2, 3].

When n = 3 and m= −1, it captures the effects of van der Waals forces in thin films [5, 9, 15].

Rewriting Eq. (3) in terms of pressure p we obtain

$$w_t = \nabla * (w^n \nabla p), \tag{4}$$

where p represents the pressure of fluid given by

$$p = -\frac{1}{m-n+1} w^{m-n+1} - \Delta w,$$

provided $n - m \neq 1$.

Suppose $\Omega \subset \mathbb{R}^2$ represents the bottom surface of a cylindrical container filled with the thin film fluid. Assuming no flux across the boundary yields the condition

$$\frac{\partial p}{\partial v} = 0 \ \ on \ \ \partial\Omega. \tag{5}$$

Additionally, neglecting wetting effects, we assume the fluid surface meets the container boundary perpendicularly, leading to

$$\frac{\partial h}{\partial v} = 0 \ \ on \ \ \partial\Omega. \tag{6}$$

Whenever $m - n \neq -1$ or -2, this system can be associated with an energy functional

$$E(w) = \int_{\Omega} \left(\frac{1}{2} |\nabla w|^2 \right)$$

and, formally using the governing equation and boundary conditions, the rate of change of energy satisfies

$$\frac{d}{dt} E(w) = - \int_{\Omega} w^n |\nabla p|^2 .$$

Thus, for a thin film fluid at rest, the pressure p must be constant, and w satisfies the elliptic equation. This framework sets the stage for exploring the asymptotic behavior of point rupture solutions, particularly in the radial setting, and their consequences for the dynamics of thin film fluids.

$$- \Delta w - \frac{1}{m-n+1} w^{m-n+1} = p \quad \text{in } \Omega,$$

For the case of Van der Waals, $m - n + 1 < -1$, the equation can be rewritten as

$$\Delta w = -\frac{1}{\alpha} w^{\alpha} - p \text{ in } \Omega, \tag{7}$$

p being an unknown constant yet to be determined and

$$\alpha = m - n + 1 > 1.$$

Recalling that $\alpha = 3$ in the case of van der Waals force driven thin film.

Therefore, (1) relate to the stationary thin film equation with van der Waals force.

Assume the elliptic equation for thin films involves the function f expressed as $f(h) = -\frac{1}{\alpha} h^{\alpha} - p$ with $\alpha \geq 0$. Therefore we are dealing with a second-order differential equation.

$$h'' + \frac{1}{r} h' + \frac{1}{\alpha} h^{\alpha} + p = 0. \tag{8}$$

Considering re-scaling assume $h^{p,\eta}$ to be the unique solution to (8) satisfying $h(0) = \eta \neq (\alpha p)^{-\frac{1}{\alpha}}$. This was completely solved in [12, 13] and some smooth radial solutions are shown below (Fig. 1).

In the case when $\eta = 0$, $h^{p,0,k}$ will be the unique rupture solution to (8). See the following plots for the rupture solutions for $k = 1, 2, 3, 4, 5,$ and 6 (Fig. 2).

In a fascinating study of critical points and re-scaling in partial differential equations, we explore a sequence of functions and their intriguing behavior under Neumann boundary conditions. Consider the sequence for p, η , $k = 1, 2, \cdots$, define $(r_k^{p,\eta})$, which represents an ascending arrangement of the critical points for the function.

In a fascinating study of critical points and re-scaling in partial differential equations, we explore a sequence of functions and their intriguing behavior under

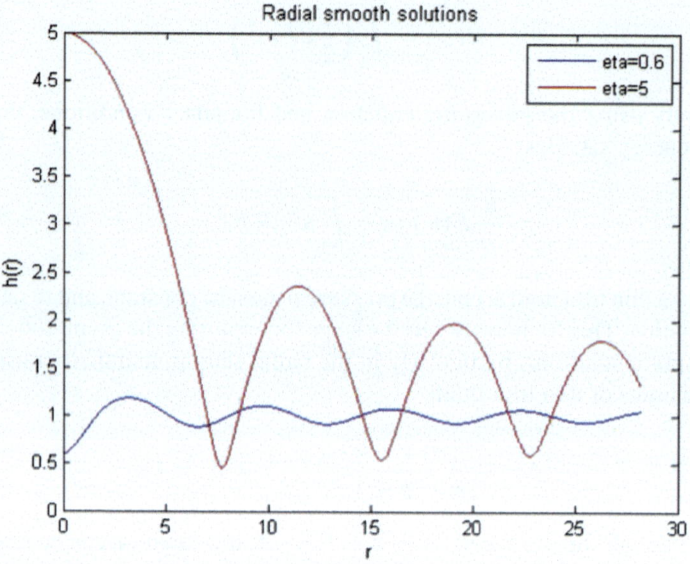

Fig. 1 Smooth radial solutions, eta positive

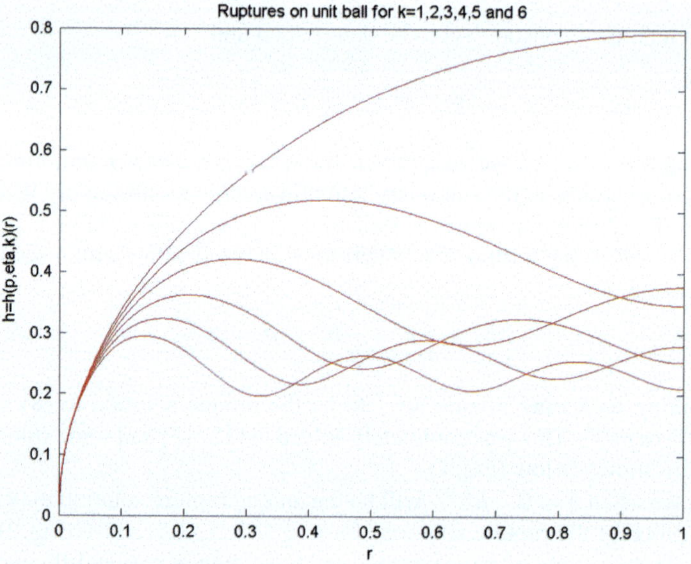

Fig. 2 Rupture solutions, Eta = 0

Neumann boundary conditions. Consider the sequence

$$h^{p,\eta,k}(x) = (r_k^{p,\eta})^{-\frac{2}{1+\alpha}} h^{p,\eta}(|x|)$$

we arrive at a formulation that satisfies a Neumann boundary value problem:

$$\begin{cases} \Delta h = \frac{1}{\alpha} \cdot h^{-\alpha} - p^{p,\eta,k} \text{ in } B_1(0), \\ \frac{\partial h}{\partial \nu} = 0 \qquad on \ \partial B_1(0), \end{cases} \tag{9}$$

where the constant is intricately defined as $p^{p,\eta,k}$ is defined as

$$p^{p,\eta,k} = p((r_k^{p,\eta})^{-\frac{2}{1+\alpha}}).$$

To better understand the role of these re-scaled functions, we define the average volume of $\bar{h}(p, \eta, k)$ by

$$\bar{h}(p, \eta, k) = \frac{1}{|B_1(0)|} \int_{B_1(0)} h^{p,\eta,k}(x)dx = \frac{(r_k^{p,\eta})^{-\frac{1}{2}}}{|B_1(0)|} \int_{B_{r_k}(0)} h^{p,\eta}(x)dx.$$

Interestingly, this average can also be expressed in terms of the original function $h^{p,\eta,k}(r)$.

What makes this analysis particularly remarkable is that $h^{p,\eta,k}(r)$, the radial representation of the solution not only solves the Newman problem but does so with a fascinating symmetry: the domain Ω is effectively transformed into the unit ball, and the average volume $\bar{h}$ takes on a constant value:

$$\bar{h} = \bar{h}(p, \eta, k).$$

Delving deeper into the connection between η and p for the solution $h^{p,\eta}$, we discover a remarkable scaling relationship:

$$h^{p,\eta}(x) = (\alpha p)^{-\frac{1}{\alpha}} h^{\frac{1}{\alpha},(\alpha p)^{\frac{1}{\alpha}\eta}} \left((\alpha p)^{\frac{1+\alpha}{2\alpha}} x \right).$$

This leads to analogous relationships for the critical points and constants:

$$r_k^{p,\eta} = (\alpha p)^{-\frac{1+\alpha}{2\alpha}} r_k^{\frac{1}{\alpha},(\alpha p)^{\frac{1}{\alpha}\eta}}, \quad h^{p,\eta,k} = h^{\frac{1}{\alpha},(\alpha p)^{\frac{1}{\alpha}\eta},k}, \quad \text{and } p^{p,\eta,k} = p^{\frac{1}{\alpha},(\alpha p)^{\frac{1}{\alpha}\eta},k}$$

and furthermore the average volume is computed as follows:

$$\bar{h}(p, \eta, k) = \bar{h}\left(\frac{1}{\alpha}, (\alpha p)^{\frac{1}{\alpha}\eta}, k\right).$$

The mechanism of producing radial and rupture solutions for the above elliptic partial differential equation was derived in [10, 11]. We also conjecture that the ruptures are discrete for finite energy solutions, and expect that the radial point rupture solutions will serve as the blow up profile of the solution near any point rupture.

A key focus is on the limiting behavior of $\bar{h}(\eta, k)$, for fixed k as $\eta \to 0$, $\eta \to 1$ and $\eta \to \infty$. Additionally, as

The work takes an intriguing turn when we consider the ordinary differential equation:

$$\begin{cases} u'' + \frac{1}{x}u' = \frac{1}{3}u^{-3} - \frac{1}{3}, \\ u(0) = \eta, \quad \text{and} \quad u'(0) = 0. \end{cases} \tag{10}$$

We will define the function z by re-scaling as $\eta z(\beta x) = u(x)$ and then get the following regular perturbation problem that we would like to study as ϵ tends to zero from above:

$$\begin{cases} z'' + \frac{1}{\xi}z' = \frac{\epsilon^3}{z^3} - 1. \\ z(0) = 1, \quad \text{and} \quad z'(0) = 0, \end{cases} \tag{11}$$

where we define $\epsilon = \frac{1}{\eta}$ and the re-scaled space variable $\xi = \beta x$ with $\beta = \frac{1}{\sqrt{3\eta}}$.

2 Main Results

In this section, we state and prove results of the asymptotic solution.

Theorem 1 *For every $\epsilon > 0$, let $z(x, \epsilon)$ be the solution of the initial value problem.*

$$\begin{cases} z'' + \frac{1}{x}z' = \frac{\epsilon^3}{z^3} - 1. \\ z(0) = 1, \quad \textit{and} \quad z'(0) = 0. \end{cases} \tag{1}$$

Then as ϵ tends to zero positively $z(x, \epsilon)$ converges uniformly to $z_*(x)$, the solution of the limiting IVP:

$$\begin{cases} z_*'' + \frac{1}{x}z_*' = -1, \quad z_* > 0 \ in \bigcup_{j=0}^{\infty}(a_j, a_{j+1}). \\ z_*(0) = 1, \quad and \ z_*'(0) = 0, \\ z_*(a_j) = 0, \quad z_*'(a_j-) = \sqrt{E(j)}, \quad z_*'(a_j+) = -\sqrt{E(j)} \end{cases}, \tag{2}$$

where $a_0 = 0$, $a_1 < a_2 < \cdots$ are inductively computed by solving the IVP (2).

Proof For any $\delta > 0$ there exists an $\epsilon > 0$ such that $x_m \geq a(\epsilon) \geq 2 - \delta$, see Appendix 4 for the proof this claim.

So consider $x \in [0, \delta]$ the solution z will satisfy the IVP:

$$\begin{cases} z'' + \frac{1}{x}z' = \frac{\epsilon^3}{z^3} - 1. \\ z(0) = 1, \quad \text{and} \quad z'(0) = 0, \end{cases} \tag{3}$$

and the asymptotic solution will satisfy the limiting IVP

$$\begin{cases} z_*'' + \frac{1}{x} z_*' = -1, . \\ z_*(0) = 1, \quad \text{and} \quad z_*'(0) = 0, \end{cases} \tag{4}$$

recall that both the solutions z_* and z must satisfy

$$z_*(x) \geq 1 - \frac{x^2}{4} \quad \text{and since } x \in [0, \delta], \quad z_*(x) \geq 1 - \frac{(2-\delta)^2}{4}.$$

Let's define the function $w(x) = z(x) - z_*(x)$, we observe that $w(x) \geq 0$ for x and hence $z \geq z_*$, the function w satisfies the IVP:

$$\begin{cases} w'' + \frac{1}{x} w' = \frac{\epsilon^3}{z^3}. \\ w(0) = 0, \quad \text{and} \quad w'(0) = 0. \end{cases} \tag{5}$$

Therefore,

$$w'' + \frac{1}{x} w' \leq \frac{\epsilon^3}{1 - \frac{(2-\delta)^2}{4}} \quad \text{or} \quad w'' + \frac{1}{x} w' \leq \frac{4\epsilon^3}{\delta(2-\delta)}.$$

Hence,

$$(xw')' \leq \frac{4x\epsilon^3}{\delta(2-\delta)} \Rightarrow w(x) \leq \frac{x^2 x\epsilon^3}{\delta(2-\delta)} \leq \frac{\epsilon^3(2-\delta)}{\delta}$$

we also have

$$w'(x) \leq \frac{2\epsilon^3}{\delta} \quad \text{and} \quad w''(x) \leq \frac{2\epsilon^3(4-\delta)}{\delta(2-\delta)}.$$

Therefore we have uniform convergence of the solution z and its first and second derivatives on the interval $[0, 2-\delta]$.

Let us now look at the inductive construction of the asymptotic solution.

$$\text{In } [0, a_1], \ (xz_*')' = -x \ \text{ so, } xz_*' = -\frac{x^2}{2} \ \text{ thus } z_*' = -\frac{x}{2} \ \text{ therefore, } z_* = 1 - \frac{x^2}{4}$$

$$a_1 = 2 \ \text{ and } E(1) = |z_*'(a_1)|^2 = 1.$$

$$\text{In } [a_1, a_2], \ (xz_*')' = -x, \ \ z_*'(2+) = 1 \ \text{ so, } xz_*' = 4 - \frac{x^2}{2} \ \text{ thus } z_*' = \frac{4}{x} - \frac{x}{2}$$

$$\text{therefore, } z_*(x) = 4\, ln\left(\frac{x}{2}\right) + \frac{4-x^2}{4}.$$

Note that $z''_* = -\frac{4}{x^2} - \frac{1}{2} < 0$ which make sense since it must be concave, and clearly there exists a unique $a_2 > 2\sqrt{2}$ such that $z_*(a_2) = 0$ in fact $a_2 = 3.74853$ and $E(2) = |z'_*(a_2)| = \frac{a_2}{2} - \frac{4}{a_2} = 0.807177$

$$\text{In } [a_j, a_{j+1}],\ (xz'_*)' = -x,\ \ z'_*(a_j+) = \sqrt{E(j)} = \frac{c_j^2}{2a_j} - \frac{a_j}{2} \ \text{ so,}$$

$$z_*(x) = \frac{c_j^2}{2}\, ln\left(\frac{x}{a_j}\right) - \frac{x^2 - a_j^2}{4}$$

the value of a_{j+1} is computed by solving the equation:

$$(a_j^2 + 2a_j\sqrt{E(j)})\ ln\left(\frac{x}{a_j}\right) + \frac{a_j^2 - x^2}{2} = 0, \quad \text{and } x > |c_j| > a_j.$$

Now the energy at a_{j+1} will be computed as

$$\sqrt{E(j+1)} = -z'_*(a_{j+1}) = -\frac{c_j^2}{2a_{j+1}} + \frac{a_{j+1}}{2} = \frac{a_{j+1}}{2} - \frac{c_j^2}{2a_{j+1}}$$

therefore we have

$$c_{j+1}^2 = 2a_{j+1}^2 - c_j^2.$$

The following is a plot for the solution with eta = 100 as an approximation to the asymptotic solution (Fig. 3).

2.1 Rupture Solution as a Limit of Smooth Solutions $\eta \to 0$

As $\eta \to 0^+$, h_η converges uniformly to the rupture solution h_0 on $[0, \infty)$. Hence, $\bar{h}(\eta, k)$ is continuous at $\eta = 0$. See the following plots for illustration (Fig. 4).

Inspired by the numerical suggestions, we would like to intend to prove that for fixed η the average volume $\bar{h}^{\eta,k}$ tends to zero as $k \to \infty$ in a decreasing manner. We will prove that indeed $\bar{h}^{\eta,k} \to 0$ as $k \to \infty$, however we are unable to prove monotonicity here.

Proposition 1 *For any fixed $\eta \geq 0$ then $\bar{h}^{\eta,k} \to 0$ as $k \to \infty$.*

Proof We know that as $k \to \infty$ the critical point $r_k^\eta \to \infty$. We also know that as $r \to \infty$ the radial solution $h(r)$ tends to 1. Therefore for any $R > 0$ there exists $N > 0$ such that for $k > N$ then $r_k^\eta > R$ and for some $R > 0$ the function $h(r) < 2$. Thus since we can write

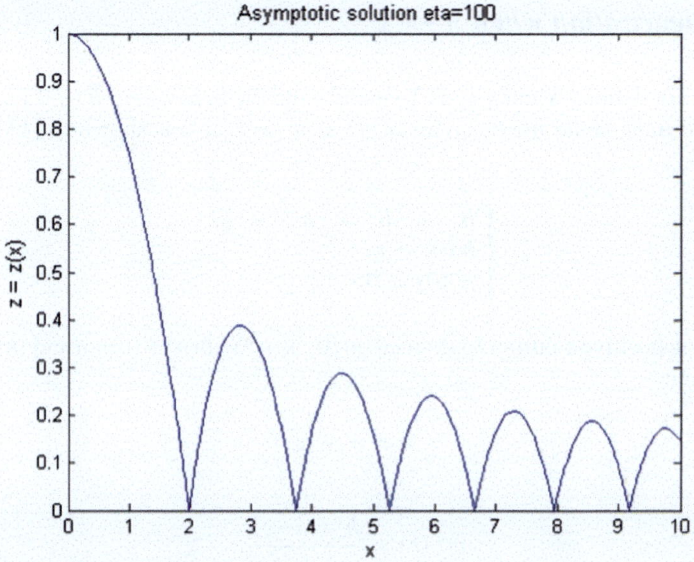

Fig. 3 Plot for asymptotic solution eta = 100

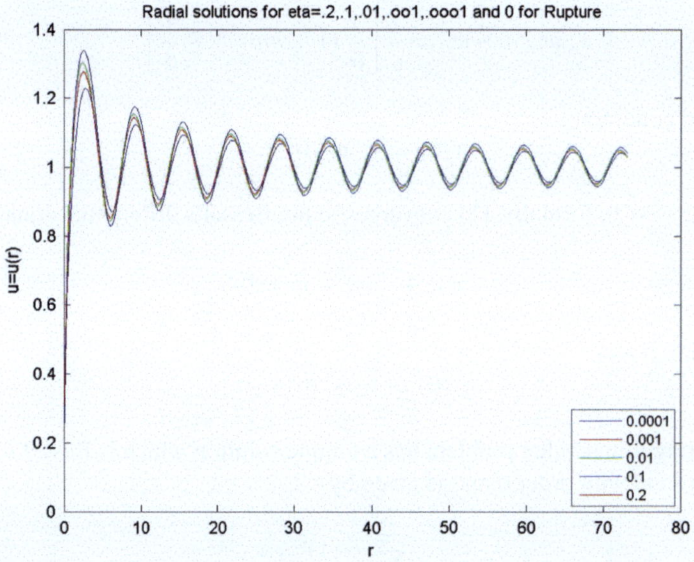

Fig. 4 Rupture solution as limit of radial smooth solutions

$$\bar{h}^{\eta,k} = 2(r_k^\eta)^{-\frac{2}{1+\alpha}-2}\left(\int_0^R rh^\eta(r)dr + \int_R^{r_k^\eta} rh^\eta(r)dr\right),$$

then the result follows.

2.2 Linearization when $\eta \to 1$

$\bar{h}(\eta, k)$ is not defined when $\eta = 1$. To understand the behavior of $\bar{h}(\eta, k)$ as $\eta \to 1$, we need to understand the behavior of h^η as $\eta \to 1$. It was shown in [11] that for any $\eta > 0$:

$$\begin{cases} h_{rr} + \frac{1}{r}h_r = \frac{1}{\alpha}h^{-\alpha} - \frac{1}{\alpha}, \\ h(0) = \eta, \\ h'(0) = 0 \end{cases} \tag{6}$$

has a unique positive solution h^η defined on $[0, \infty)$ /5with $h^\eta(0) = \eta$ and $(h^\eta)'(0) = 0$. We define

$$\varepsilon = \eta - 1,$$

and

$$w^\eta(r) = \frac{h^\eta(r) - 1}{\varepsilon}.$$

Then w^η is a solution to the differential equation

$$w_{rr} + \frac{1}{r}w_r = \frac{1}{\varepsilon}\left[\frac{1}{\alpha}(1 + \varepsilon w)^{-\alpha} - \frac{1}{\alpha}\right] \tag{7}$$

with initial condition

$$w(0) = 1, \; w'(0) = 0.$$

As $\eta \to 1$, $\varepsilon \to 0$, formally, (7) converges to the Bessel's differential equation with order 0:

$$w_{rr} + \frac{1}{r}w_r + w = 0$$

with the initial date

$$w(0) = 1, \; w'(0) = 0.$$

Such limiting initial value problem has a unique solution which is Bessel's function of the first kind with order 0 and is given by

$$J_0(x) = \sum_{n=0}^{\infty} \frac{(-1)^n}{(n!)^2}\left(\frac{x}{2}\right)^{2n}.$$

We remark here that $J_0(x)$ is oscillating around 0. See plot of Bessel function of the first kind of order zero in the next page.

We can show that as $\eta \to 1$, w^η converges uniformly to J_0. Since both w^η and J_0 are oscillating around 0, $r_k^\eta \to r_k^*$ as $\eta \to 1$ where r_k^*, $k = 1, 2, \cdots$ is an increasing sequence of the critical radius of Bessel's function $J_0(x)$. Since $h^\eta \to 1$ uniformly as $\eta \to 1$, we have

$$\bar{h}^{\eta,k} = \frac{(r_k^\eta)^{-\frac{2}{1+\alpha}}}{|B_{r_k^\eta}(0)|}\int_{B_{r_k^\eta}(0)} h^\eta(r)dr \to (r_k^*)^{-\frac{2}{1+\alpha}}$$

as $\eta \to 1$ and

$$\begin{aligned} E^{\eta,k} &= (r_k^\eta)^{-\frac{4}{1+\alpha}}\int_{B_{r_k^\eta}(0)}\left(\frac{1}{2}|\nabla h^\eta|^2 - \frac{1}{\alpha(\alpha-1)}(h^\eta)^{1-\alpha}\right) \\ &\to -\frac{1}{\alpha(\alpha-1)}(r_k^*)^{-\frac{4}{1+\alpha}}\left|B_{r_k^*}(0)\right| = -\frac{\pi}{\alpha(\alpha-1)}(r_k^*)^{2-\frac{4}{1+\alpha}} \end{aligned}$$

as $\eta \to 1$.

Hence, we can define $\bar{h}^{\eta,k}$ and $E^{\eta,k}$ so that they are both continuous functions on $[0,\infty)$.

2.3 *Limiting Profile when* $\eta \to \infty$

In this section, we want to understand the behavior of $\bar{h}^{\eta,k}$ and $E^{\eta,k}$ as $\eta \to \infty$.

Let $\eta > 1$ and h_η be the solution to (6). We define the blow down solution z by

$$z(x) = \frac{1}{\eta}h(r)$$

with $r = \sqrt{\alpha\eta}x$. Then we have

$$\begin{aligned} z_x &= \frac{\sqrt{\alpha}}{\sqrt{\eta}}h_r(r), \\ z_{xx} &= \alpha h_{rr}(r) \end{aligned}$$

and hence

$$\begin{aligned} z'' + \frac{1}{x}z' &= \alpha\left(h_{rr} + \frac{1}{r}h_r\right) = h^{-\alpha} - 1 \\ &= \frac{\eta^{-\alpha}}{z^\alpha} - 1. \end{aligned}$$

Denoting $\varepsilon = \frac{1}{\eta}$, we have $\varepsilon \to 0$ as $\eta \to \infty$. The blow down function z is a solution to the initial value problem:

$$\begin{cases} z'' + \frac{1}{x}z' = \frac{\varepsilon^\alpha}{z^\alpha} - 1, \\ z(0) = 1, \text{ and } z'(0) = 0. \end{cases} \tag{8}$$

The following plot illustrates the blow down function z for $\varepsilon = 10$.
Formally, as $\varepsilon \to 0$, (8) converges to the limiting equation:

$$\begin{cases} z'' + \frac{1}{x}z' = -1, \\ z(0) = 1, \text{ and } z'(0) = 0, \end{cases} \tag{9}$$

which has a unique global solution

$$z(x) = 1 - \frac{1}{4}x^2.$$

However, we can't expect

$$\lim_{\varepsilon \to 0} z^\varepsilon(x) = 1 - \frac{1}{4}x^2$$

since the function $1 - \frac{1}{4}x^2$ becomes negative when $x > 2$.

The following theorem follows:

Theorem 2 *For every $\varepsilon > 0$, let $z^\varepsilon(x)$ be the unique solution of the initial value problem* (8).

Then as ε tends to zero positively, $z^\varepsilon(x)$ converges uniformly on $[0, \infty)$ to $z_*(x)$, the solution of the limiting initial value problem:

$$\begin{cases} z_*'' + \frac{1}{x}z_*' = -1, \quad z_* > 0 \;\; in \bigcup_{j=0}^{\infty}(a_j, a_{j+1}). \\ z_*(0) = 1, \quad and \;\; z_*'(0) = 0, \\ z_*(a_j) = 0, \quad z_*'(a_j+) = -z_*'(a_j-) \end{cases}, \tag{10}$$

where $a_0 = 0$, $2 = a_1 < a_2 < \cdots$ are inductively computed by solving the IVP (10).

Hence, in particular, we have $z^\varepsilon(x)$ converges uniformly to $1 - \frac{1}{4}x^2$ on $[0, 2]$ as $\varepsilon \to 0$ and $\frac{r_1^\eta}{\sqrt{\alpha\eta}}$ converges to 2 as $\eta \to \infty$. More generally, we have for $k = 1, 2, 3, \cdots$,

$$\lim_{\eta \to \infty} \frac{r_{2k-1}^\eta}{\sqrt{\alpha\eta}} = a_k$$

and

$$\lim_{\eta \to \infty} \frac{r_{2k}^\eta}{\sqrt{\alpha\eta}} = b_k,$$

where b_k is the maximum point of z^* in (a_k, a_{k+1}).

Given a positive integer k and given $\eta > 1$, we have

$$\bar{h}^{\eta,k} = 2(r_k^\eta)^{-\frac{2}{1+\alpha}-2} \int_0^{r_k^\eta} r h^\eta(r) dr$$

$$= 2(r_k^\eta)^{-\frac{2}{1+\alpha}-2}\eta \int_0^{r_k^\eta} rz\left(\frac{r}{\sqrt{\alpha\eta}}\right) dr$$

$$= 2(r_k^\eta)^{-\frac{2}{1+\alpha}-2}\alpha\eta^2 \int_0^{\frac{r_k^\eta}{\sqrt{\alpha\eta}}} sz\,(s)\,ds$$

$$= 2\alpha^{-\frac{1}{1+\alpha}}\eta^{\frac{\alpha}{1+\alpha}}\left(\frac{r_k^\eta}{\sqrt{\alpha\eta}}\right)^{-\frac{2}{1+\alpha}-2} \int_0^{\frac{r_k^\eta}{\sqrt{\alpha\eta}}} sz\,(s)\,ds.$$

Hence, we have for $k = 1, 2, 3, \cdots$,

$$\lim_{\eta\to\infty} \frac{\bar{h}^{\eta,2k-1}}{\eta^{\frac{\alpha}{1+\alpha}}} = 2\alpha^{-\frac{1}{1+\alpha}} a_k^{-\frac{2}{1+\alpha}-2} \int_0^{a_k} sz^*\,(s)\,ds$$

and

$$\lim_{\eta\to\infty} \frac{\bar{h}^{\eta,2k}}{\eta^{\frac{\alpha}{1+\alpha}}} = 2\alpha^{-\frac{1}{1+\alpha}} b_k^{-\frac{2}{1+\alpha}-2} \int_0^{b_k} sz^*\,(s)\,ds.$$

We remark here that for each positive integer k, $\bar{h}^{\eta,k} \to \infty$ as $\eta \to \infty$.

The following plots illustrate how the average volume $\bar{h}$ of the fluid changes as η changes from 0 to 25 and k = 1, 2, 3 (Fig. 5).

Next we investigate the energy of radial solutions as $\eta \to \infty$. Since

$$z\,(x) = \frac{1}{\eta}h(r),$$

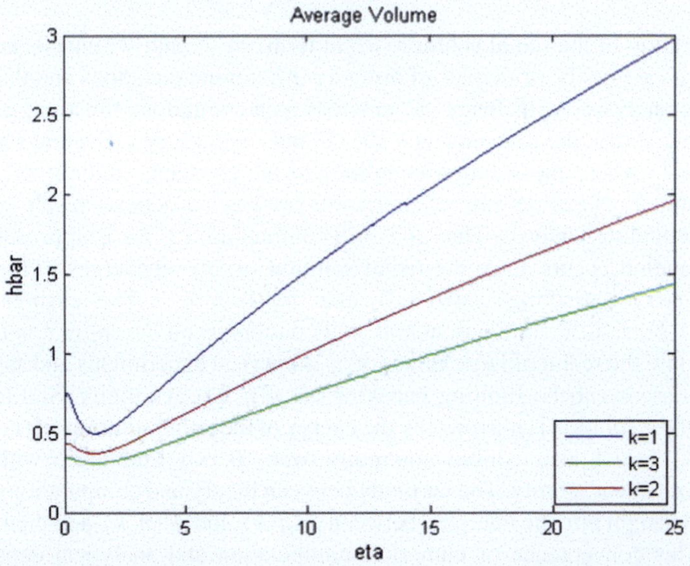

Fig. 5 Eta in [0, 25], k = 1, 2, 3

$$\frac{dh^{\eta}}{dr}=\eta z'(x)\frac{dx}{dr}=\frac{\sqrt{\eta}}{\sqrt{\alpha}}z'(x).$$

Hence

$$\begin{aligned}E^{\eta,k}&=2\pi(r_k^{\eta})^{-\frac{4}{1+\alpha}}\int_0^{r_k^{\eta}}\left(\frac{1}{2}\left(\frac{dh^{\eta}}{dr}\right)^2-\frac{1}{\alpha(\alpha-1)}(h^{\eta})^{1-\alpha}\right)rdr\\&=2\pi(\frac{r_k^{\eta}}{\sqrt{\alpha\eta}})^{-\frac{4}{1+\alpha}}(\alpha\eta)^{1-\frac{2}{1+\alpha}}\int_0^{\frac{r_k^{\eta}}{\sqrt{\alpha\eta}}}\left(\frac{1}{2}\frac{\eta}{\alpha}(z'(x))^2-\frac{1}{\alpha(\alpha-1)}(\eta z(x))^{1-\alpha}\right)xdx\\&=\pi\alpha^{-\frac{2}{1+\alpha}}(\frac{r_k^{\eta}}{\sqrt{\alpha\eta}})^{-\frac{4}{1+\alpha}}\eta^{2-\frac{2}{1+\alpha}}\int_0^{\frac{r_k^{\eta}}{\sqrt{\alpha\eta}}}x\left|z'\right|^2dx+O\left(\eta^{2-\alpha-\frac{2}{1+\alpha}}\right).\end{aligned}$$

Hence, we have for $k=1,2,3,\cdots$,

$$\lim_{\eta\to\infty}\frac{E^{\eta,2k-1}}{\eta^{2-\frac{2}{1+\alpha}}}=\pi\alpha^{-\frac{2}{1+\alpha}}a_k^{-\frac{4}{1+\alpha}}\int_0^{a_k}s\left|(z^*)'\right|^2ds$$

and

$$\lim_{\eta\to\infty}\frac{E^{\eta,2k}}{\eta^{2-\frac{2}{1+\alpha}}}=\pi\alpha^{-\frac{2}{1+\alpha}}b_k^{-\frac{4}{1+\alpha}}\int_0^{b_k}s\left|(z^*)'\right|^2ds.$$

3 Conclusion

The exploration of the radial solution $\bar{u}$, in terms of η and k unveils an elegant narrative shaped by the principles of ordinary differential equations and the rigor of asymptotic analysis. At its heart, $\bar{u}$ emerges as a continuous function, gracefully transitioning across the domains $\eta\in[0,1]$ and $\eta\in[1,\infty)$. Even at the critical point $\eta=1$ where the solution simplifies to the constant solution $u\equiv1$, the critical point r_1 loses definition, continuity persists—a delicate result confirmed through meticulous analysis. Thus $\bar{u}$ is still continuous at 1. As η approaches zero, a transformation occurs since the radial solution $u_{\eta}(r)$ converges to the rupture solution $u_0(r)$ satisfying $u_0(0)=0$ and $u_0'(0)=\infty$. This ensures that for every fixed k , $\bar{u}(\eta,k)$ remains smooth and continuous on the entire range $[0,\infty)$. Venturing into the realm of large values of η numerical experiments and asymptotic investigations reveal the limiting behavior of $\bar{u}(\eta,k)$ at infinity. Similarly, this behavior of the solution is mirrored by the energy of the radial solution u_{η}^{k}(r) denoted by $E(\eta,k)$ which also exhibits continuity over $[0,\infty)$ and aligns with a well-defined profile near infinity. The harmony between theory and computation provides a profound insight into the interplay between $\bar{u}(\eta,k)$ and $E(\eta,k)$ and their limiting profiles. This convergence of numerical computation and analytical depth paints a vivid picture of how radial solutions and their energies gracefully evolve as η changes, offering a richer understanding of their intricate relationships.

4 Appendix

Lemma 1 *For any $\delta > 0$ there is an $\epsilon > 0$ such that $a(\epsilon) \geq 2 - \delta$.*

Proof We have

$$z'' + \frac{1}{x}z' = \frac{\epsilon^3}{z^3} - 1 \quad \text{thus} \quad z'' + \frac{1}{x}z' \geq -1 \quad \text{therefore} \quad xz'' + z' \geq -x$$

$$(xz')' \geq -x \ \text{ integrating we get, } \int_0^x (yz'(y))'dy \geq \int_0^x -ydy \ \text{ that is, } \ xz'(x) \geq -\frac{x^2}{2}$$

dividing both sides by x and integrating a second time, we obtain

$$\int_0^x z'dy \geq \int_0^x \frac{-y}{2}dy \quad \text{that is,} \quad z(x) - 1 \geq -\frac{x^2}{4}$$

this holds for any x; in particular, when $x = a(\epsilon)$ we get

$$\epsilon - 1 \geq -\frac{a^2(\epsilon)}{4} \quad \text{thus} \ a^2(\epsilon) \geq 4(1-\epsilon) \quad \text{therefore} \quad a(\epsilon) \geq 2\sqrt{1-\epsilon}.$$

So given any $\delta > 0$ we will need an $\epsilon > 0$ such that $2\sqrt{1-\epsilon} \geq 2 - \delta$ and this will occur when $0 \leq \epsilon \leq 1 - \frac{(2-\delta)^2}{4}$ and for such ϵ we will have $x_m \geq a(\epsilon) \geq 2 - \delta$ as required.

References

1. Almgren, R., Bertozzi, A., Brenner, M.P.: Stable and unstable singularities in the unforced Hele-Shaw cell. Phys. Fluids **8**(6), 1356–1370 (1996)
2. Bernoff, A.J., Bertozzi, A.L.: Singularities in a modified Kuramoto-Sivashinsky equation describing interface motion for phase transition. Phys. D **85**(3), 375–404 (1995)
3. Bertozzi, A.L., Pugh, M.C.: Finite-time blow-up of solutions of some long-wave unstable thin film equations. Indiana Univ. Math. J. **49**(4), 1323–1366 (2000)
4. Bertozzi, A.L., Brenner, M.P., Dupont, T.F., Kadanoff, L.P.: Singularities and similarities in interface flows. In: Trends and Perspectives in Applied Mathematics. Applied Mathematical Sciences, vol. 100, pp. 155–208. Springer, New York (1994)
5. Chen, X., Jiang, H.: Singular limit of an energy minimizer arising from dewetting thin film model with van der waal, born repulsion and surface tension forces. Calc. Var. Partial Differ. Equ. **44**(1–2), 221–246 (2012)
6. Constantin, P., Dupont, T.F., Goldstein, R.E., Kadanoff, L.P., Shelley, M.J., Zhou, S.-M.: Droplet breakup in a model of the Hele-Shaw cell. Phys. Rev. E (3), **47**(6), 4169–4181 (1993)
7. Dupont, T.F., Goldstein, R.E., Kadanoff, L.P., Zhou, S.-M.: Finite-time singularity formation in Hele-Shaw systems. Phys. Rev. E (3), **47**(6), 4182–4196 (1993)
8. Ehrhard, P.: The spreading of hanging drops. J. Colloid Interface **168**(1), 242–246 (1994)
9. Jiang, H., Lin, F.: Zero set of soblev functions with negative power of integrability. Chinese Ann. Math. Ser. B **25**(1), 65–72 (2004)

10. Jiang, H., Miloua, A.: Point rupture solutions of a singular elliptic equation. Electron. J. Differ. Eq. **2013**(70), 1–8 (2013)
11. Jiang, H., Ni, W.-M.: On steady states of van der Waals force driven thin film equations. Eur. J. Appl. Math **18**(2), 153–180 (2007)
12. Laugesen, R.S., Pugh, M.C.: Properties of steady states for thin film equations. Eur. J. Appl. Math **11**(3), 293–351 (2000)
13. Shelley, M.J., Goldstein R.E., Pesci, A.I.: Topological transitions in HeleShaw flow. In: Singularities in Fluids, Plasmas and Optics (Heraklion, 1992), NATO Advanced Science Institute Series C: Mathematical and Physical Sciences, vol. 404, pp. 167–188. Kluwer Academic Publisher, Dordrecht (1993)
14. Witelski, T.P., Bernoff, A.J.: Stability of self-similar solutions for van der Waals driven thin film rupture. Phys. Fluids **11**(9), 2443–2445 (1999)
15. Zhang, W.W., Lister, J.R.: Similarity solutions for van der Waals rupture of a thin film on a solid substrate. Phys. Fluids **11**(9), 2454–2462 (1999)

Uniform Approximation of Function $g \in L[0, \infty)$-Space Using $E^q.T$-Means of its Fourier–Laguerre Series

Shailesh Kumar Srivastava, Sachin Devaiya, and Lakshmi Narayan Mishra

Abstract This study determines the error of approximation of function g belonging to the $L\,[0, \infty)$-class by the $E^q.T$-means of its Fourier–Laguerre series for any $y > 0$, and further discusses certain special cases of $E^q.T$-means.

Keywords Error of approximation · $E^q.T$-mean · Fourier–Laguerre series

1 Introduction

Let g be a Lebesgue integrable function on $L[0, \infty)$-space, i.e.: $g \in L[0, \infty)$. The Fourier–Laguerre expansion of function $g \in L[0, \infty)$ is given by

$$g(y) \sim \sum_{m=0}^{\infty} a_m\, \mathcal{L}_m^{(\alpha)}(y), \tag{1}$$

where mth Laguerre polynomial, $\mathcal{L}_m^{(\alpha)}(y)$, of order $0 < \alpha + 1$, is defined by the generating function:

$$\sum_{m=0}^{\infty} \mathcal{L}_m^{(\alpha)}(y)\, \omega^m = \frac{e^{\left(\frac{\omega y}{\omega - 1}\right)}}{(1-\omega)^{1+\alpha}},$$

S. K. Srivastava · S. Devaiya (✉)
Department of Mathematics, Sardar Vallabhbhai National Institute of Technology, Surat, Gujarat, India
e-mail: sbdevaiya18695@gmail.com

L. N. Mishra
Department of Mathematics, School of Advanced Sciences, Vellore Institute of Technology, Vellore, Tamil Nadu, India

R. K. Sharma et al. (eds.), *Mathematical Theory and Simulation of Scientific Problems*, Springer Proceedings in Mathematics & Statistics 487,
https://doi.org/10.1007/978-981-96-2579-6_6

and

$$a_m = \int_0^{\infty} \frac{m!\,\alpha!\,g(y)\,y^{\alpha}\,\mathcal{L}_m^{(\alpha)}(y)}{(m+\alpha)!\,\Gamma(1+\alpha)\,e^y}dy. \tag{2}$$

It is assumed that the above integral exists.

The $(m+1)$th partial sum of the Fourier–Laguerre series (1) is defined by

$$s_m(g;\,y) = \sum_{b=0}^{m} a_b\,\mathcal{L}_b^{(\alpha)}(y),\ m \in \mathrm{N} \cup \{0\}. \tag{3}$$

It is also known as Fourier–Laguerre polynomial of error (or order) $\leq (m+1)$.

Let $T \equiv (a_{m,b})$ be an infinite lower triangular matrix with real or complex entries such that $A_{m,b} = \sum_{r=b}^{m} a_{m,r}$. Then the sequence-to-sequence transformation

$$[t]_m(g;\,y) = \sum_{b=0}^{m} a_{m,b}\,s_b(g;\,y),\ m \in \mathrm{N} \cup \{0\}$$

defines the matrix transform of the sequence $\{s_m(g;\,y)\}_{m=0}^{\infty}$ generated by the elements $a_{m,d}$ of the matrix T, and we call it the matrix means of the Fourier–Laguerre series (1). The Fourier–Laguerre series is called T-summable to the sum s, if $[t]_m(g;\,y) \to s$ as $m \to \infty$.

If $a_{m,b} = \begin{cases} (q+1)^{-m}\binom{m}{b}q^{m-b}, & 0 \leq b \leq \text{m}, \\ 0, & b > \text{m}, \end{cases}$ then the matrix T reduces to Euler summability and denoted by E^q. The Fourier–Laguerre series is said to be E^q-summable $(q > 0)$ to the sum s, if $[E^q]_m(g;\,y) \to s$ as $m \to \infty$.

$E^q.T$-summable is the product of T-summability and E^q-summability. $[E^q.T]_m(g;\,y)$ denotes the $E^q.T$-summability of sequence $\{s_m(g;\,y)\}$, and defined by

$$[E^q.T]_m(g;\,y) = (q+1)^{-m}\sum_{c=0}^{m}\binom{m}{c}q^{m-c}\sum_{b=0}^{c} a_{c,b}s_b(g;\,y). \tag{4}$$

The Fourier–Laguerre series is called $E^q.T$-summable to the sum s, if $[E^q.T]_m(g;\,y) \to s$ as $m \to \infty$. If T and E^q are regular then $E^q.T$ is also regular.

The essential and particular cases of the $E^q.T$ method are listed below:

1. If $a_{c,b} = \frac{p_{c-b}\,q_b}{R_c}$ where $R_c = \sum_{b=0}^{c} p_b\,q_{c-b}$, then $E^q.T$ reduces to $(E, q)(N, p, q)$ or $E^q.N_{pq}$.
2. If $a_{c,b} = \frac{c!\,\eta!}{(c+\eta)!}\binom{c+\eta-b-1}{\eta-1}$, then $E^q.T$ reduces to $(E, q)(C, \eta)$ or $E^q.C^{\eta}$.
3. If $a_{c,b} = \frac{p_{c-b}}{P_c}$ where $P_c = \sum_{b=0}^{c} p_b \neq 0$, then $E^q.T$ reduces to $(E, q)(N, p_c)$ or $E^q.N_p$.
4. If $a_{c,b} = \frac{p_b}{P_c}$, then $E^q.T$ reduces to $(E, q)(\overline{N}, p_c)$ or $E^q.\overline{N}_p$.
5. If $a_{c,b} = \frac{1}{log(c+1)\,(c-b+1)}$, then $E^q.T$ reduces to $(E, q)(H, \frac{1}{c+1})$ or $E^q.H$.

If we take $q = 1$ in the above cases, then we get

6. If $a_{c,b} = \frac{p_{c-b}\, q_b}{R_c}$ where $R_c = \sum_{b=0}^{c} p_b\, q_{c-b}$, then $E^1.T$ reduces to $(E, 1)(N, p, q)$ or $E^1.N_{pq}$.
7. If $a_{c,b} = \frac{c!\,\eta!}{(c+\eta)!}\binom{c+\eta-b-1}{\eta-1}$, then $E^1.T$ reduces to $E^1.C^{\eta}$ or $(E, 1)(C, \eta)$.
8. If $a_{c,b} = \frac{p_{c-b}}{P_c}$ where $P_c = \sum_{b=0}^{c} p_b \neq 0$, then $E^1.T$ reduces to $E^1.N_p$ or $(E, 1)(N, p_c)$.
9. If $a_{c,b} = \frac{p_b}{P_c}$, then $E^1.T$ reduces to $E^1.\overline{N}_p$ or $(E, 1)(\overline{N}, p_c)$.
10. If $a_{c,b} = \frac{1}{log(c+1)\,(c-b+1)}$, then $E^1.T$ reduces to $E^1.H$ or $(E, 1)(H, \frac{1}{c+1})$.

Here p_c and q_c are monotonic, non-increasing, and non-negative sequence of real constants.

A function g belonging to $L[0, \infty)$-space is approximated by a Fourier–Laguerre polynomials $T_m(y)$ of error $\leq (m+1)$ (which are either partial sums or some summability means of the Fourier–Laguerre series of g), and the error of approximation $E_m(g)$, in terms of m, is given by

$$E_m(g) = \min_{T_m(y)} \operatorname*{ess\,sup}_{y \in [0,\infty)} |T_m(y) - g(y)|.$$

The kernel polynomial, denoted as $\mathcal{J}_b^{\alpha}(y, z)$, is given as

$$\mathcal{J}_b^{\alpha}(y, z) = \sum_{m=0}^{b} \frac{\mathcal{L}_m^{(\alpha)}(y)\, \mathcal{L}_m^{(\alpha)}(z)}{\Gamma(1+\alpha)\binom{m+\alpha}{m}}, \tag{5}$$

which can also be written as

$$\mathcal{J}_b^{\alpha}(y, z) = \frac{b+1}{\Gamma(1+\alpha)\binom{\alpha+b}{\alpha}} \frac{\mathcal{L}_b^{(\alpha)}(y)\, \mathcal{L}_{b+1}^{(\alpha)}(z) - \mathcal{L}_{b+1}^{(\alpha)}(y)\, \mathcal{L}_b^{(\alpha)}(z)}{y - z}. \tag{6}$$

In more appropriate form, we can write

$$\mathcal{J}_b^{\alpha}(y, z) = \frac{b+1}{\Gamma(1+\alpha)\binom{\alpha+b}{\alpha}} \frac{\mathcal{L}_{b+1}^{(\alpha)}(y)\, \mathcal{L}_{b+1}^{(\alpha-1)}(z) - \mathcal{L}_{b+1}^{(\alpha-1)}(y)\, \mathcal{L}_{b+1}^{(\alpha)}(z)}{y - z}, \tag{7}$$

and

$$\mathcal{J}_b^{\alpha}(y, z) = \frac{b+1}{\Gamma(1+\alpha)\binom{\alpha+b}{\alpha}} \Bigg(\mathcal{L}_{b+1}^{(\alpha)}(y) \frac{\mathcal{L}_{b+1}^{(\alpha-1)}(z) - \mathcal{L}_{b+1}^{(\alpha-1)}(y)}{y - z} - \mathcal{L}_{b+1}^{(\alpha-1)}(y) \frac{\mathcal{L}_{b+1}^{(\alpha)}(z) - \mathcal{L}_{b+1}^{(\alpha)}(y)}{y - z} \Bigg). \tag{8}$$

More detailed information is available in [11, pp. 101, 266].

Remark 1 We consider the series $1 - 4046\sum_{m=1}^{\infty}(-4045)^{m-1}$. The mth partial sum of the series is given by $s_m = (-4045)^m$.

Here, we take matrix $T \equiv (a_{m,b})$, where $a_{m,b} = \begin{cases} \frac{1}{2023^m}\binom{m}{b}(2022)^{m-b}, & 0 \le b \le \mathrm{m}, \\ 0, & b > m, \end{cases}$ then from the definition of T-summability and E^q-summability, we can say that the series is neither T-summable nor E^q-summable, but it is $E^q.T$-summable. As a result, it is obvious that product summability are better than the single summability.

Numerous investigators [1–7, 10] have approximated the function $g \in L[0, \infty)$-class using various summability methods of its Fourier–Laguerre series at a point $y = 0$. The authors like Singh [8], and Singh and Saini [9] have approximated function $g \in L[0, \infty)$ by Cesàro means of the Fourier–Laguerre series of g for any $y > 0$.

Throughout this paper, we use following notations:

$\psi(y, u) = g(y \pm u) - g(y)$ and $\phi(y, z) = g(z) - g(y)$.

2 Lemmas

Here, we are giving some lemmas which are useful for proving our theorems:

Lemma 1 ([11, pp. 177, Theorem 7.6.4]) *Let $\epsilon > 0$ and $\alpha \in \mathrm{R}$. Then*

$$\mathcal{L}_m^{(\alpha)}(y) = \begin{cases} O(m^{\alpha}), & 0 \le y \le 1/m, \\ O(y^{-(21+\alpha)/4}\, m^{(2\alpha-1)/4}), & 1/m \le y \le \epsilon, \end{cases} \quad \text{as } m \to \infty.$$

When $\alpha + 1/2 \ge 0$, the bounds are satisfied in both intervals.

Lemma 2 ([11, pp. 241, Theorem 8.91.7]) *Let $\alpha \in \mathrm{R}$, $\rho \in \mathrm{R}$, $\eta \in (0, 4)$ and $0 < \omega$. Then*

$$max\, \frac{y^{\rho}|\mathcal{L}_m^{(\alpha)}(y)|}{e^{y/2}} = O(m^{Q}),$$

where

$$Q = \begin{cases} max\left(\frac{\alpha}{2} - \frac{1}{4}, \rho - \frac{1}{2}\right), & \omega \le y \le (4-\eta)m, \\ max\left(\frac{\alpha}{2} - \frac{1}{4}, \rho - \frac{1}{3}\right), & y > m. \end{cases}$$

Lemma 3 ([11, pp. 198, Theorem 8.22.1]) *Let $\epsilon > 0$, $\omega > 0$ and $\alpha \in \mathrm{R}$, then*

$$\mathcal{L}_m^{(\alpha)}(y) = \frac{e^{y/2}\cos(2\sqrt{my} - (1+\alpha/2)\pi/2)}{\sqrt{\pi}\, m^{1/4-\alpha/2}\, y^{\alpha/2+1/4}} + O(m^{\alpha/2-3/4}), \text{ for } y \in [\epsilon, \omega].$$

Lemma 4 ([11, pp. 237]) *If $y \in [1/m, \omega]$ and $z \in [1/m, \omega]$, then*

$$\frac{\mathcal{L}_m^{(\alpha)}(z) - \mathcal{L}_m^{(\alpha)}(y)}{\sqrt{z} - \sqrt{y}} = \frac{e^{z/2}\cos(2\sqrt{mz} + \gamma) - \cos(2\sqrt{my} + \gamma)}{m^{1/4-\alpha/2}\, z^{\alpha/2+1/4}(\sqrt{z} - \sqrt{y})} +$$

$$\frac{O(m^{(2\alpha-1)/4})}{y^{(2\alpha+3)/4}} + \frac{O(m^{(2\alpha-1)/4})}{z^{(2\alpha+3)/4}},$$

where $\gamma = -(1 + \alpha/2)\pi/2$.

Lemma 5 ([9, Lemma 5]) *If* $\Phi(t) = \int_t^{\epsilon} \frac{|\phi(y,z)|}{z^{1/4-\alpha/2}} dz = o\left(\xi\left(\frac{1}{t}\right)\right)$, *for* $t \to 0$ *holds, then*

$$\int_0^t z^{\alpha} |\phi(y, z)| \, dz = o\left(t^{\alpha/2+1/4} \xi\left(\frac{1}{t}\right)\right).$$

Lemma 6 ([9, Lemma 6]) *If* $\int_m^{\infty} \frac{e^{-z/2} |\phi(y,z)|}{z^{13/12-\alpha/2}} dz = o\left(\frac{\xi(m)}{m^{1/2}}\right)$, $m \to \infty$, *holds, then*

$$\int_{\omega}^{m} \frac{z^{(2\alpha-3)/4} |\phi(y, z)|}{e^{z/2}} dz = o(\xi(m)),$$

where $m \to \infty$ *and* ω *is a fixed positive number.*

3 Main Results

Numerous efforts have been made in approximating function $g \in L[0, \infty)$-space through various summability methods of its Fourier–Laguerre series at the point $y = 0$. However, there needs to be more exploration in the context of $y > 0$. Remark 1 shows that product summability method is powerful than the single summability method. Considering the above-mentioned specific cases of $E^q.T$-means and Remark 1, we approximate the function g belonging to the $L[0, \infty)$-space by $E^q.T$ for their Fourier–Laguerre series for any $y > 0$.

Theorem 1 *Let* $T \equiv (a_{m,b})$ *be a lower triangular regular matrix which satisfy the following conditions:*

1. $a_{m,b} \geq 0$, *and* $a_{m,b} \leq a_{m,b+1}$, *for* $0 \leq b \leq m$,
2. $a_{m,-1} = 0$, *and* $A_{m,0} = 1$, $m \in \mathrm{N} \cup \{0\}$,
3. $\sum_{c=t}^{m} A_{c,c-t} = O(m + 1)$, $m \in \mathrm{N} \cup \{0\}$.

Thus, for any $y > 0$, *the error of approximation of function* g *belonging to* $L[0, \infty)$-*class by* $E^q.T$-*means of its Fourier–Laguerre series is given by*

$$\left|[E^q.T]_m(g; y) - g(y)\right| = o(\xi(m)), \tag{9}$$

where $\xi(t)$ *denotes a positive, increasing function with the property* $\xi(t) \to \infty$ *as* $t \to \infty$, *and it satisfies the conditions below:*

$$\Phi(t) = \int_t^{\epsilon} \frac{|\phi(y, z)|}{z^{1/4-\alpha/2}} dz = o\left(\xi\left(\frac{1}{t}\right)\right), \quad t \to 0, \tag{10}$$

$$\int_t^\delta \frac{|\psi(y,u)|}{u}du = o\left(\xi\left(\frac{1}{t}\right)\right), \quad t \to 0, \tag{11}$$

$$\int_m^\infty \frac{e^{-z/2}\,|\phi(y,z)|}{z^{13/12-\alpha/2}}dz = o\left(\frac{\xi(m)}{m^{1/2}}\right), \quad m \to \infty, \tag{12}$$

where $\alpha + 1/2 \geq 0$ and δ is a fixed positive constant. This remains true uniformly for every fixed positive interval $0 < \epsilon \leq y \leq \omega < \infty$.

Proof We have

$$\begin{aligned} s_m(g;y) &= \sum_{b=0}^{m} a_b\,\mathcal{L}_b^{(\alpha)}(y) \\ &= \sum_{b=0}^{m} \frac{\int_0^\infty z^\alpha\, e^{-z}\, g(z)\,\mathcal{L}_b^{(\alpha)}(z)\,\mathcal{L}_b^{(\alpha)}(y)\,dz}{\binom{\alpha+b}{\alpha}\,\Gamma(\alpha+1)} \\ &= \frac{1}{\Gamma(\alpha+1)}\int_0^\infty z^\alpha\, e^{-z}\, g(z) \sum_{b=0}^{m} \frac{\mathcal{L}_b^{(\alpha)}(y)\,\mathcal{L}_b^{(\alpha)}(z)}{\binom{\alpha+b}{\alpha}}dz \\ &= \int_0^\infty z^\alpha\, e^{-z}\, g(z)\,\mathcal{J}_m^\alpha(y,z)\,dz. \end{aligned} \tag{13}$$

On Eq. (13), by applying T-summability, we obtain the following expression:

$$\begin{aligned} [T]_m(g;y) &= \sum_{b=0}^{m} a_{m,b}\, s_b(y) \\ &= \sum_{b=0}^{m} a_{m,b} \int_0^\infty z^\alpha\, e^{-z}\, g(z)\,\mathcal{J}_b^\alpha(y,z)\,dz. \end{aligned} \tag{14}$$

On Eq. (14), by applying (E,q)-summability, we obtain the following expression:

$$\begin{aligned} [E^q.T]_m(g;y) = \frac{(1+q)^{-m}}{\Gamma(1+\alpha)} &\sum_{c=0}^{m} \binom{m}{c} q^{m-c} \sum_{b=0}^{c} a_{c,c-b} \times \\ &\int_0^\infty z^\alpha\, e^{-z}\, g(z)\,\mathcal{J}_b^\alpha(y,z)\,dz, \end{aligned} \tag{15}$$

we have

$$\begin{aligned} &[E^q.T]_m(g;y) - g(y) \\ &= (q+1)^{-m} \sum_{c=0}^{m} \binom{m}{c} q^{m-c} \sum_{b=0}^{c} a_{c,c-b} \int_0^\infty e^{-z} z^\alpha\, \phi(y,z)\,\mathcal{J}_b^\alpha(y,z)\,dz \end{aligned}$$

$$= (q+1)^{-m} \sum_{c=0}^{m} \binom{m}{c} q^{m-c} \sum_{b=0}^{c} a_{c,c-b} \left[\int_0^{1/m} + \int_{1/m}^{\epsilon} + \int_{\epsilon}^{y-\delta} + \int_{y-\delta}^{y+\delta} + \int_{y+\delta}^{\omega} + \int_{\omega}^{m} + \int_{m}^{\infty} \right] z^{\alpha} e^{-z} \phi(y,z) \mathcal{J}_b^{\alpha}(y,z)\, dz$$

$$= \sum_{j=1}^{7} M_j. \tag{16}$$

Considering y as a specific positive constant, and based on the second part of Lemma 1, we obtain

$$\begin{aligned} |\mathcal{L}_{b+1}^{(\alpha)}(y)| &= O(b^{(2\alpha-1)/4}\ y^{(-2\alpha+1)/4}) \\ &= O(b^{(2\alpha-1)/4}). \end{aligned} \tag{17}$$

By applying Lemma 1 to Eq. (7) for $z \in [0, 1/m)$, we obtain

$$|\mathcal{J}_b^{\alpha}(y,z)| = O(b^{1-\alpha}[b^{\alpha-1}\, b^{\alpha/2-1/4} + b^{\alpha}\, b^{\alpha/2-3/4}]). \tag{18}$$

Using Lemma 5 and Eqs. (17), (18), we have

$$|M_1| \le (q+1)^{-m} \sum_{c=0}^{m} \binom{m}{c} q^{m-c} \sum_{b=0}^{c} a_{c,c-b} \int_0^{1/m} z^{\alpha} |\phi(y,z)|\, |\mathcal{J}_b^{\alpha}(y,z)|\, dz$$

$$\begin{aligned} &= (q+1)^{-m} \sum_{c=0}^{m} \binom{m}{c} q^{m-c} \sum_{b=0}^{c} a_{c,c-b}\, O\left(b^{1-\alpha} \int_0^{1/m} z^{\alpha} |\phi(y,z)| \right. \\ &\qquad \left. [b^{\alpha/2-3/4} b^{\alpha} + b^{\alpha/2-1/4} b^{\alpha-1}]\, dz\right) \\ &= (q+1)^{-m} \sum_{c=0}^{m} \binom{m}{c} q^{m-c} \sum_{b=0}^{c} a_{c,c-b}\, O\left(b^{\alpha/2+1/4} \int_0^{1/m} z^{\alpha} |\phi(y,z)| dz\right) \\ &= (q+1)^{-m} \sum_{c=0}^{m} \binom{m}{c} q^{m-c} \sum_{b=0}^{c} a_{c,c-b}\, o(\xi(m)) \\ &= o\left(\xi(m)\, (q+1)^{-m} \sum_{c=0}^{m} \binom{m}{c} q^{m-c}\right) \\ &= o(\xi(m)). \end{aligned} \tag{19}$$

Now, applying Lemma 1 within Eq. (7) for the interval $1/m \le z < \epsilon$, we derive the following:

$$|\mathcal{J}_b^\alpha(y,z)| = O(b^{1-\alpha}[b^{\alpha/2-1/4}z^{-\alpha/2+1/4}b^{\alpha/2-3/4} + b^{\alpha/2-3/4}z^{-\alpha/2-1/4}b^{\alpha/2-1/4}]). \tag{20}$$

Using condition (10) and Eq. (20), we have

$$\begin{aligned}
|M_2| &\le (q+1)^{-m}\sum_{c=0}^{m}\binom{m}{c}q^{m-c}\sum_{b=0}^{c}a_{c,c-b}\int_{1/m}^{\epsilon} z^\alpha\,|\phi(y,z)|\,|\mathcal{J}_b^\alpha(y,z)|\,dz\\
&= (q+1)^{-m}\sum_{c=0}^{m}\binom{m}{c}q^{m-c}\sum_{b=0}^{c}a_{c,c-b}\,O\left(b^{1-\alpha}\int_{1/m}^{\epsilon} z^\alpha\,|\phi(y,z)|\right.\\
&\qquad \left.[b^{\alpha/2-1/4}z^{-\alpha/2+1/4}b^{\alpha/2-3/4} + b^{\alpha/2-3/4}z^{-\alpha/2-1/4}b^{\alpha/2-1/4}]\,dz\right)\\
&= (q+1)^{-m}\sum_{c=0}^{m}\binom{m}{c}q^{m-c}\sum_{b=0}^{c}a_{c,c-b}\,O\left(\int_{1/m}^{\epsilon} z^{\alpha/2-1/4}\,|\phi(y,z)|dz\right)\\
&= o\left(\xi(m)\,(q+1)^{-m}\sum_{c=0}^{m}\binom{m}{c}q^{m-c}\right)\\
&= o(\xi(m)).
\end{aligned} \tag{21}$$

By applying Lemma 3 in the interval $\epsilon \le z \le y-\delta$, we derive the following:

$$|\mathcal{L}_{b+1}^{(\alpha-1)}(z)| = O\left(\frac{e^{z/2}\cos(2\sqrt{bz}-(\alpha-1/2)\pi/2)}{\sqrt{\pi}\,z^{\alpha/2-1/4}\,b^{3/4-\alpha/2}} + b^{\alpha/2-5/4}\right). \tag{22}$$

Using formula (7), we obtain

$$\begin{aligned}
M_3 &= (q+1)^{-m}\sum_{c=0}^{m}\binom{m}{c}q^{m-c}\sum_{b=0}^{c}a_{c,c-b}\int_{\epsilon}^{y-\delta} z^\alpha\,e^{-z}\\
&\qquad \phi(y,z)\mathcal{J}_b^\alpha(y,z)dz\\
&= (q+1)^{-m}\sum_{c=0}^{m}\binom{m}{c}q^{m-c}\sum_{b=0}^{c}a_{c,c-b}O(b^{1-\alpha})\int_{\epsilon}^{y-\delta} z^\alpha\,e^{-z}\,\phi(y,z)\\
&\qquad \frac{\mathcal{L}_{b+1}^{(\alpha)}(y)\,\mathcal{L}_{b+1}^{(\alpha-1)}(z) - \mathcal{L}_{b+1}^{(\alpha-1)}(y)\,\mathcal{L}_{b+1}^{(\alpha)}(z)}{y-z}dz\\
&= M_{31} + M_{32}.
\end{aligned} \tag{23}$$

Using Eq. (22) and Lemma 3, we have

$$\begin{aligned}
M_{31} &= (q+1)^{-m}\sum_{c=0}^{m}\binom{m}{c}q^{m-c}\sum_{b=0}^{c}a_{c,c-b}O(b^{1-\alpha})\int_{\epsilon}^{y-\delta}\\
&\qquad \frac{z^\alpha\,e^{-z}\,|\phi(y,z)|}{y-z}O(b^{\alpha/2-1/4})\left[\frac{b^{\alpha/2-3/4}\,e^{z/2}}{\sqrt{\pi}\,z^{\alpha/2-1/4}}\right.
\end{aligned}$$

$$
\begin{aligned}
&\cos(\sqrt{4bz} - (2\alpha - 1)\pi/4) + O(b^{(2\alpha-5)/4})\Big] dz \\
&= (q+1)^{-m} \sum_{c=0}^{m} \binom{m}{c} q^{m-c} \sum_{b=0}^{c} a_{c,c-b} \Big[O(1) \int_{\epsilon}^{y-\delta} \\
&\frac{|\phi(y,z)|\, z^{(2\alpha+1)/4}}{(y-z)\, e^{z/2}} \cos(2\sqrt{bz} - (2\alpha - 1)\pi/4) dz + \\
&O(b^{-1/2}) \int_{\epsilon}^{y-\delta} \frac{e^{-z} z^{\alpha} |\phi(y,z)|}{y-z} dz + o(1) \Big] \qquad (24) \\
&= (q+1)^{-m} \sum_{c=0}^{m} \binom{m}{c} q^{m-c} \sum_{b=0}^{c} a_{c,c-b}\, o(1) \\
&= o\left((q+1)^{-m} \sum_{c=0}^{m} \binom{m}{c} q^{m-c} \right) \\
&= o(1), \qquad (25)
\end{aligned}
$$

in Eq. (24), the Riemann–Lebesgue theorem ensures that the first integral goes to 0, and the second integral approaches 0 as b becomes infinitely large.

Similarly, we can find

$$|M_{32}| = o(1). \tag{26}$$

Combining Eqs. (23), (25), and (26), we have

$$|M_3| = o(1). \tag{27}$$

Similarly, we can find

$$|M_5| = o(1). \tag{28}$$

Using Lemmas 3 and 4 in formula (8), we have

$$
\begin{aligned}
\mathcal{J}_b^{\alpha}(y,z) &= b^{1-\alpha}\, b^{\alpha/2-1/4}\, b^{\alpha/2-3/4} \frac{y^{-\alpha/2-1/4}\, e^{y/2} z^{-\alpha/2-1/4}\, e^{z/2}}{\pi(\sqrt{y} + \sqrt{z})} \\
&\Big[\sqrt{z} \cos(2\sqrt{by} + \gamma) \frac{\sin(2\sqrt{bz} + \gamma) - \sin(2\sqrt{by} + \gamma)}{\sqrt{z} - \sqrt{y}} - \sqrt{y} \\
&\sin(2\sqrt{by} + \gamma) \frac{\cos(2\sqrt{bz} + \gamma) - \cos(2\sqrt{by} + \gamma)}{\sqrt{z} - \sqrt{y}} + O(1) \Big] \qquad (29)
\end{aligned}
$$

For the estimation of M_4, Lemmas 3 and 4 are utilized in formula (8). Given that the variables are limited to a fixed positive interval, the remainder terms in these lemmas are dependent only on m. Consequently, the following conclusion is drawn (see [11, pp. 267]):

$$\mathcal{J}_b^{\alpha}(y,z)=\frac{1}{2}\sqrt{\frac{y}{z}}\,\frac{\sin(2\sqrt{bz}-2\sqrt{by})}{(\sqrt{z}-\sqrt{y})}\left(\frac{e^{y/2}\,y^{-\alpha/2-1/4}}{\sqrt{\pi}}\right)^2+O(1).$$

From Eq., we have

$$\begin{aligned}
M_4&=(q+1)^{-m}\sum_{c=0}^{m}\binom{m}{c}q^{m-c}\sum_{b=0}^{c}a_{c,c-b}\int_{y-\delta}^{y+\delta}e^{-z}z^{\alpha}\phi(y,z)\mathcal{J}_b^{\alpha}(y,z)dz\\
&=(q+1)^{-m}\sum_{c=0}^{m}\binom{m}{c}q^{m-c}\sum_{b=0}^{c}a_{c,c-b}\left[\frac{1}{2}\sqrt{y}\left(\frac{e^{y/2}}{\sqrt{\pi}\,y^{\alpha/2+1/4}}\right)^2\right.\\
&\left.\int_{y-\delta}^{y+\delta}e^{-z}\,z^{\alpha-1/2}\,\phi(y,z)\frac{\sin(2\sqrt{b}(\sqrt{z}-\sqrt{y}))}{\sqrt{z}-\sqrt{y}}dz+O(1)\right]\\
&=(q+1)^{-m}\sum_{c=0}^{m}\binom{m}{c}q^{m-c}\sum_{b=0}^{c}a_{c,c-b}O\left(\left[\int_{y-\delta}^{y-1/m}+\int_{y-1/m}^{y+1/m}+\right.\right.\\
&\left.\left.\int_{y+1/m}^{y+\delta}\right]e^{-z}\,z^{(2\alpha-1)/2}\phi(y,z)\,\frac{\sin(2\sqrt{b}(\sqrt{z}-\sqrt{y}))}{\sqrt{z}-\sqrt{y}}dz\right)+O(1)\\
&=M_{41}+M_{42}+M_{43}+O(1).
\end{aligned}\tag{30}$$

Applying condition (11), we have

$$\begin{aligned}
|M_{41}|&=(q+1)^{-m}\sum_{c=0}^{m}\binom{m}{c}q^{m-c}\sum_{b=0}^{c}a_{c,c-b}O\left(\int_{y-\delta}^{y-1/m}\right.\\
&\qquad\left.\frac{(\sqrt{y}+\sqrt{z})\,|\phi(y,z)|}{|y-z|}dz\right)\\
&=(q+1)^{-m}\sum_{c=0}^{m}\binom{m}{c}q^{m-c}\sum_{b=0}^{c}a_{c,c-b}O\left(\int_{1/m}^{\delta}\frac{\psi(y,u)}{u}du\right)\\
&=o\left(\xi(m)\,(q+1)^{-m}\sum_{c=0}^{m}\binom{m}{c}q^{m-c}\right)\\
&=o(\xi(m)).
\end{aligned}\tag{31}$$

Following the above proof, we have

$$|M_{43}|=o(\xi(m)).\tag{32}$$

Applying condition (11), we have

$$
\begin{aligned}
|M_{42}| &= (q+1)^{-m} \sum_{c=0}^{m} \binom{m}{c} q^{m-c} \sum_{b=0}^{c} a_{c,c-b} O\left(\int_{y-1/m}^{y+1/m} |\phi(y,z)| \right. \\
&\qquad \left. \left| \frac{\sin(2\sqrt{bz} - 2\sqrt{by})}{\sqrt{z} - \sqrt{y}} \right| dz \right) \\
&= (q+1)^{-m} \sum_{c=0}^{m} \binom{m}{c} q^{m-c} \sum_{b=0}^{c} a_{c,c-b}\, O\left(\sqrt{b} \int_0^{1/m} |\psi(y,u)| du \right) \\
&= (q+1)^{-m} \sum_{c=0}^{m} \binom{m}{c} q^{m-c} \sum_{b=0}^{c} a_{c,c-b}\, o\left(\frac{\sqrt{b}\, \xi(m)}{m} \right) \\
&= o\left(\xi(m)\, (q+1)^{-m} \sum_{c=0}^{m} \binom{m}{c} q^{m-c} \right) \\
&= o(\xi(m)). \qquad (33)
\end{aligned}
$$

Combining Eqs. (30)–(33), we have

$$
|M_4| = o(\xi(m)). \qquad (34)
$$

Using first part of Lemma 2 (for $\eta = 3$), we have

$$
\begin{aligned}
|M_6| &\leq (q+1)^{-m} \sum_{c=0}^{m} \binom{m}{c} q^{m-c} \sum_{b=0}^{c} a_{c,c-b} \left[O(b^{1-\alpha}) \int_{\omega}^{m} e^{-z} z^{\alpha-1} \right. \\
&\quad |\phi(y,z)| |\mathcal{L}_{b+1}^{(\alpha)}(y)| |\mathcal{L}_{b+1}^{(\alpha-1)}(z)| dz + O(b^{1-\alpha}) \int_{\omega}^{m} e^{-z} z^{\alpha-1} |\phi(y,z)| \\
&\quad \left. |\mathcal{L}_{b+1}^{(\alpha-1)}(y)| |\mathcal{L}_{b+1}^{(\alpha)}(z)| dz \right] \\
&= M_{61} + M_{62}. \qquad (35)
\end{aligned}
$$

Applying Lemma 6, we have

$$
\begin{aligned}
|M_{61}| &= (q+1)^{-m} \sum_{c=0}^{m} \binom{m}{c} q^{m-c} \sum_{b=0}^{c} a_{c,c-b} O\left(b^{-\alpha/2+3/4} \int_{\omega}^{m} e^{-z/2} \right. \\
&\qquad \left. z^{\alpha/2-3/4} |\phi(y,z)| e^{-z/2}\, z^{\alpha/2-1/4} |\mathcal{L}_{b+1}^{(\alpha-1)}(z)| dz \right) \\
&= (q+1)^{-m} \sum_{c=0}^{m} \binom{m}{c} q^{m-c} \sum_{b=0}^{c} a_{c,c-b} (b^{-\alpha/2+3/4}\, b^{\alpha/2-3/4}) \\
&\qquad \int_{\omega}^{m} e^{-z/2}\, z^{\alpha/2-3/4}\, |\phi(y,z)| dz
\end{aligned}
$$

$$
\begin{aligned}
&= (q+1)^{-m} \sum_{c=0}^{m} \binom{m}{c} q^{m-c} \sum_{b=0}^{c} a_{c,c-b} \left(b^{-\alpha/2+3/4}\, b^{\alpha/2-3/4}\right) o(\xi(m)) \\
&= o\left(\xi(m)\, (q+1)^{-m} \sum_{c=0}^{m} \binom{m}{c} q^{m-c} \right) \\
&= o(\xi(m)). \qquad (36)
\end{aligned}
$$

Similarly, we can calculate

$$
|M_{62}| = o(\xi(m)). \qquad (37)
$$

Combining Eqs. (35)–(37), we have

$$
|M_6| = o(\xi(m)). \qquad (38)
$$

By applying the second part of Lemma 2 for $m \leq y < \infty$, we have

$$
e^{-y/2}\, y^{\alpha/2+1/12}\, |\mathcal{L}_m^{(\alpha)}(y)| = O(m^{\alpha/2-1/4}). \qquad (39)
$$

Applying formula (6), we have

$$
\begin{aligned}
|M_7| &= (q+1)^{-m} \sum_{c=0}^{m} \binom{m}{c} q^{m-c} \sum_{b=0}^{c} a_{c,c-b} O(b^{1-\alpha}) \\
&\quad \int_m^{\infty} \frac{z^{\alpha}\, e^{-z} |\phi(y,z)|}{y-z} \{\mathcal{L}_b^{(\alpha)}(y)\, \mathcal{L}_{b+1}^{(\alpha)}(z) - \mathcal{L}_{b+1}^{(\alpha)}(y)\, \mathcal{L}_b^{(\alpha)}(z)\} dz \\
&= M_{71} + M_{72}. \qquad (40)
\end{aligned}
$$

Using Eqs. (12) and (39), we get

$$
\begin{aligned}
M_{71} &= (q+1)^{-m} \sum_{c=0}^{m} \binom{m}{c} q^{m-c} \sum_{b=0}^{c} a_{c,c-b} O(b^{1-\alpha}) \\
&\quad \int_m^{\infty} \frac{z^{\alpha}\, e^{-z} |\phi(y,z)|}{z-y} O(b^{\alpha/2-1/4}) e^{z/2} z^{-\alpha/2-1/12} b^{\alpha/2-1/4} dz \\
&= (q+1)^{-m} \sum_{c=0}^{m} \binom{m}{c} q^{m-c} \sum_{b=0}^{c} a_{c,c-b} O(b^{1/2}) \int_m^{\infty} e^{-z/2} \\
&\qquad\qquad z^{\alpha/2-13/12} |\phi(y,z)| dz \\
&= (q+1)^{-m} \sum_{c=0}^{m} \binom{m}{c} q^{m-c} \sum_{b=0}^{c} a_{c,c-b}\, o(\xi(m)) \\
&= o\left(\xi(m)\, (q+1)^{-m} \sum_{c=0}^{m} \binom{m}{c} q^{m-c} \right) \\
&= o(\xi(m)). \qquad (41)
\end{aligned}
$$

Similarly, we can calculate

$$|M_{72}| = o(\xi(m)). \tag{42}$$

Combining Eqs. (40)–(42), we have

$$|M_7| = o(\xi(m)). \tag{43}$$

Collecting Eqs. (16), (19), (21), (27), (28), (34), (38), and (43), we have

$$\left|[E^q.T]_m(g; y) - g(y)\right| = o(\xi(m)).$$

4 Corollaries

From Theorem 1, we derive the following corollaries.

Corollary 1 *If we consider* $a_{c,b} = \frac{p_{c-b} q_b}{R_c}$ *in (4), then* $E^q.T$ *reduces to* $(E, q)(N, p, q)$ *or* $E^q.N_{pq}$*, then for* $g \in L[0, \infty)$

$$\left|[E^q.N_{pq}]_m(g; y) - g(y)\right| = o(\xi(m)).$$

Corollary 2 *If we consider* $a_{c,b} = \frac{c!\,\eta!}{(c+\eta)!}\binom{c+\eta-b-1}{\eta-1}$ *in (4), then* $E^q.T$ *reduces to* $(E, q)(C, \eta)$ *or* $E^q.C^\eta$*, then for* $g \in L[0, \infty)$

$$\left|[E^q.C^\eta]_m(g; y) - g(y)\right| = o(\xi(m)).$$

Corollary 3 *If we consider* $a_{c,b} = \frac{p_b}{P_c}$ *in (4), then* $E^q.T$ *reduces to* $(E, q)(\overline{N}, p_c)$ *or* $E^q.\overline{N}_p$*, then for* $g \in L[0, \infty)$

$$\left|[E^q.\overline{N}_p]_m(g; y) - g(y)\right| = o(\xi(m)).$$

Corollary 4 *If we consider* $a_{c,b} = \frac{1}{log(c+1)\,(c-b+1)}$ *in (4), then* $E^q.T$ *reduces to* $(E, q)(H, \frac{1}{c+1})$ *or* $E^q.H$*, then for* $g \in L[0, \infty)$

$$\left|[E^q.H]_m(g; y) - g(y)\right| = o(\xi(m)).$$

Corollary 5 *If we consider* $a_{c,b} = \frac{p_{c-b}}{P_c}$*, where* $P_c = \sum_{b=0}^{c} p_b \neq 0$ *in (4), then* $E^q.T$ *reduces to* $(E, q)(N, p_c)$ *or* $E^q.N_p$*, then for* $g \in L[0, \infty)$

$$\left|[E^q.N_p]_m(g; y) - g(y)\right| = o(\xi(m)).$$

Remark 2 If we consider $q = 1$ in (E, q), then from above corollaries $(E, q)(N, p, q)$, $(E, q)(C, \eta)$, $(E, q)(\overline{N}, p_c)$, $(E, q)(H, \frac{1}{c+1})$, and $(E, q)(N, p_c)$ are reduced to $(E, 1)(N, p, q)$, $(E, 1)(C, \eta)$, $(E, 1)(\overline{N}, p_c)$, $(E, 1)(H, \frac{1}{c+1})$, and $(E, 1)(N, p_c)$. These are also special cases of Theorem 1.

Acknowledgements Shailesh Kumar Srivastava thanks Sardar Vallabhbhai National Institute of Technology, Surat-395007, Gujarat, India for the Grant No. 2020-21/Seed Money/26, and Sachin Devaiya thanks the Council of Scientific and Industrial Research (CSIR), New Delhi, India for the Award No. 09/1007(0008)/2020-EMR-I.

References

1. Khatri, K., Mishra, V.N.: Approximation of functions belonging to $L[0, \infty)$ by product summability means of its Fourier-Laguerre series. Cogent Math. **3**(1), 1250854 (2016). https://doi.org/10.1080/23311835.2016.1250854
2. Krasniqi, X.Z.: On the degree of approximation of a function by $(C, 1)(E, q)$ means of its Fourier-Laguerre series. Int. J. Anal. Appl. **1**(1), 33–39 (2013)
3. Lal, S., Nigam, H.K.: Degree of approximation by (N, p, q) summability means of the Fourier-Laguerre expansion. Tamkang J. Math. **32**(2), 143–150 (2001). https://doi.org/10.5556/j.tkjm.32.2001.357
4. Mittal, M.L., Singh, M.V.: Error estimation of functions by Fourier-Laguerre polynomials using Matrix-Euler operators. Int. J. Anal. (2015). https://doi.org/10.1155/2015/478345
5. Nigam, H.K., Sharma, A.: A study on degree of approximation by $(E, 1)$ summability means of the Fourier-Laguerre expansion. Int. J. Math. Math. Sci. (2010). https://doi.org/10.1155/2010/351016
6. Sahani, S.K., Mishra, V.N., Pahari, N.P.: On the degree of approximation of a function by Nörlund means of its Fourier-Laguerre series. Nepal J. Math. Sci. **1**, 65–70 (2020). https://doi.org/10.3126/njmathsci.v1i0.34164
7. Saini, S., Singh, U.: Degree of approximation of $g \in L[0, \infty)$ by means of Fourier-Laguerre series. In: Agrawal, P. N., Mohapatra, R. N., Singh, U., Srivastava, H. M. (eds.) Mathematical Analysis and its Applications. Springer Proceedings in Mathematics & Statistics, vol. 143, pp. 207–217. Springer India, New Delhi (2015). https://doi.org/10.1007/978-81-322-2485-3_16
8. Singh, T.: On the absolute summability factors of Fourier-Laguerre expansion. Indian J. Pure Appl. Math. **7**(9), 961–968 (1976)
9. Singh, U., Saini, S.: Uniform approximation in $L[0, \infty)$-space by Cesàro means of Fourier-Laguerre series. Proc. Natl. Acad. Sci., India, Sect. A Phys. Sci. (2021). https://doi.org/10.1007/s40010-021-00747-8
10. Sonker, S.: Approximation of functions by $(C^1.T)$ means of its Fourier-Laguerre series. In: Proceeding of ICMS-2014, vol. 1, no. 1, pp. 122–125 (2014)
11. Szegö, G.: Orthogonal Polynomials. American Mathematical Society, New York (1959)

The Lie Group Analysis of the 2D Time-Independent Isotropic Harmonic Oscillator

Sach Mulchan, Sreedhara Rao Gunakala, B. Rushi Kumar, Vikash Ramcharitar, and Victor M. Job

Abstract The Lie group analysis of the two-dimensional, time-independent, isotropic quantum harmonic oscillator is performed. Firstly, a simplified version of the equation called the base case equation is considered and its symmetry Lie algebra was found to be isomorphic to the Euclidean Lie algebra. Thus the symmetries that the base case equation possessed were rotations about the origin and translations in the x- and y-directions. Secondly, a new basis for the Lie algebra of the 2D harmonic oscillator was found and the equation was shown to possess rotational symmetries about the origin. Next, the base case equation was written in complex variables and its symmetries were used to obtain two group invariant solutions which were in turn used to form a general solution. Lastly, the two-dimensional harmonic oscillator was converted to complex form and its rotational symmetry was used to construct a general solution.

Keywords Euclidean Lie algebra · Isotropic harmonic oscillator · Time independent · Lie group · Rotational symmetry

1 Introduction

The relationship between group theory and the theory of differential equations was first established by Norwegian mathematician Sophus Lie during the years 1872–1899. Just as Galois generalized the theory of algebraic equations using group theory,

S. Mulchan · S. R. Gunakala (✉) · V. Ramcharitar
Department of Mathematics and Statistics, The University of the West Indies, St. Augustine, Trinidad and Tobago
e-mail: Sreedhara.Rao@sta.uwi.edu

B. R. Kumar
Department of Mathematics, School of Advanced Sciences, Vellore Institute of Technology, Vellore, Tamil Nadu, India

V. M. Job
Department of Mathematics, The University of the West Indies, Mona, Jamaica

R. K. Sharma et al. (eds.), *Mathematical Theory and Simulation of Scientific Problems*, Springer Proceedings in Mathematics & Statistics 487,
https://doi.org/10.1007/978-981-96-2579-6_7

Lie showed that a similar relationship existed between groups and differential equations. The major difference was that the groups associated with differential equations (Lie groups) are uncountably infinite. Lie groups are particularly important to the study of quantum mechanics. For example, three of the four fundamental forces of the universe can be described using Lie groups. The fourth force, gravity, is described using concepts from differential geometry which are closely related to concepts in Lie groups. Since 1899 many mathematicians and physicists have contributed to the development of this theory. The Lie algebras for both equations were determined along with bases for each algebra. The process was similar to that used in Boyer et al. [2]. The commutation relations of the basis vectors were determined.

The fundamental algebraic and analytical systems used in the formulation of the theory of Lie group methods for the solution of differential equations are Lie groups, matrix groups, Lie algebras, and manifolds. The general linear group and convergence of a sequence of matrices and matrix Lie group is as used in Hall [4]. Since a Lie group is also a manifold, the theory of Lie group methods for differential equations is based significantly on results from both differential geometry and group theory. The dimensionality of a Lie algebra is the dimensionality of the associated vector space. For example, it can easily be verified that the space of all $n \times n$ matrices with complex entries, $M_n(\mathbb{C})$, forms a Lie algebra over $\mathbb{C}$ where the usual matrix and scalar multiplication hold as given in Bluman and Anco [1]. Since a Lie algebra homomorphism preserves the commutation relations of the elements of a Lie algebra, we expect that there exists a relationship between two isomorphic Lie algebras and their structural constants [7]. There is a fundamental relationship between Lie algebras and matrix Lie groups. To highlight this relationship, we analyze the symmetries of the 2D Helmholtz equation (base case) and the 2D simple harmonic oscillator equation.

2 Development of Symmetries

2.1 The Base Case Equation

The 2D Helmholtz equation is given by

$$\frac{\partial^2 \psi}{\partial x^2} + \frac{\partial^2 \psi}{\partial y^2} + \omega^2 \psi = 0. \tag{1}$$

We consider only those solutions ψ of (1) which are defined and analytic in real variables x and y for some common open connected set $D \subseteq \mathbb{R}^2$.

The set of all such solutions forms a complex vector space $\mathcal{V}_0$. Setting D as fixed, $\mathcal{V}_0$ is called the solution space of (1). Let $\mathcal{V}$ be the vector space of all complex-valued, real analytic functions defined on D and let $\mathbf{H}$ be the partial differential operator

$$\mathbf{H} = \frac{\partial^2}{\partial x^2} + \frac{\partial^2}{\partial y^2} + \omega^2 \tag{2}$$

defined on D. Thus, if $\phi \in \mathcal{V}$, then $\mathbf{H}\phi \in \mathcal{V}$ and equation (1) can be written as $\mathbf{H}\psi(x, y) = 0$. This implies that $\mathcal{V}_0$ is the kernel or null space of the linear operator $\mathbf{H}$, which is a subspace of $\mathcal{V}$.

2.2 *The 2D Simple Harmonic Oscillator*

Consider the 2D simple harmonic oscillator:

$$\frac{-h^2}{2M}\left(\frac{\partial^2\psi}{\partial x^2} + \frac{\partial^2\psi}{\partial y^2}\right) + \frac{1}{2}M\omega(x^2 + y^2)\psi = E\psi. \tag{3}$$

The Hamiltonian for this equation is given by

$$\mathbf{H} = \alpha\left(\frac{\partial^2}{\partial x^2} + \frac{\partial^2}{\partial y^2}\right) + \beta(x^2 + y^2), \tag{4}$$

where $\alpha = \dfrac{-h^2}{2M}$ and $\beta = \dfrac{1}{2}M\omega^2$.

To obtain the symmetries of the equation we consider only those solutions ψ which are defined and analytic in real variables x and y for some common open, connected set $D \subseteq \mathbb{R}^2$ such that

$$\alpha\left(\frac{\partial^2\psi}{\partial x^2} + \frac{\partial^2\psi}{\partial y^2}\right) + \beta(x^2 + y^2)\psi = 0. \tag{5}$$

The set of all such solutions forms a complex vector space $\mathcal{V}_0$. Setting D as fixed, $\mathcal{V}_0$ is called the solution space of (5). Let $\mathcal{V}$ be the vector space of all complex-valued, real analytic functions defined on D. Thus, if $\phi \in \mathcal{V}$, then $\mathbf{H}\phi \in \mathcal{V}$. Therefore, $\mathcal{V}_0$ is the null space of the linear operator $\mathbf{H}$. We use the symmetry operator $\mathbf{X}$ and explicitly determine the symmetry algebra, for $f \in \mathcal{V}$:

$$\begin{aligned}
\mathbf{HX}f &= \left[\alpha\left(\frac{\partial^2}{\partial x^2} + \frac{\partial^2}{\partial y^2}\right) + \beta\left(x^2 + y^2\right)\right]\left[\xi\frac{\partial f}{\partial x} + \eta\frac{\partial f}{\partial y} + \tau f\right] \\
&= \alpha\xi\frac{\partial^3 f}{\partial x^3} + 2\alpha\frac{\partial \xi}{\partial x}\frac{\partial^2 f}{\partial x^2} + \alpha\frac{\partial f}{\partial x}\frac{\partial^2 \xi}{\partial x^2} + \alpha\eta\frac{\partial^3 f}{\partial x^2 \partial y} + 2\alpha\frac{\partial \eta}{\partial x}\frac{\partial^2 f}{\partial x \partial y} \\
&\quad + \alpha\frac{\partial f}{\partial y}\frac{\partial^2 \eta}{\partial x^2} + \alpha\tau\frac{\partial^2 f}{\partial x^2} + 2\alpha\frac{\partial \tau}{\partial x}\frac{\partial f}{\partial x} + \alpha\frac{\partial^2 \tau}{\partial x^2}f \\
&\quad + \alpha\xi\frac{\partial^3 f}{\partial x \partial y^2} + 2\alpha\frac{\partial \xi}{\partial y}\frac{\partial^2 f}{\partial x \partial y} + \alpha\frac{\partial^2 \xi}{\partial y^2}\frac{\partial f}{\partial x} + \alpha\eta\frac{\partial^3 f}{\partial y^3}
\end{aligned}$$

$$
\begin{aligned}
&+ 2\alpha \frac{\partial \eta}{\partial y}\frac{\partial^2 f}{\partial y^2} + \alpha \frac{\partial f}{\partial y}\frac{\partial^2 \eta}{\partial y^2} + \alpha\tau \frac{\partial^2 f}{\partial y^2} + 2\alpha \frac{\partial \tau}{\partial y}\frac{\partial f}{\partial y} + \alpha \frac{\partial^2 \tau}{\partial y^2} f \\
&+ \beta\left(x^2+y^2\right)\left[\xi \frac{\partial f}{\partial x} + \eta \frac{\partial f}{\partial y} + \tau f\right] \\
&= \xi\left[\alpha\left(\frac{\partial^3 f}{\partial x^3} + \frac{\partial^3 f}{\partial x \partial y^2}\right) + \beta\left(x^2+y^2\right)\frac{\partial f}{\partial x}\right] \\
&+ \eta\left[\alpha\left(\frac{\partial^3 f}{\partial x^2 \partial y} + \frac{\partial^3 f}{\partial y^3}\right) + \beta\left(x^2+y^2\right)\frac{\partial f}{\partial y}\right] \\
&+ \tau\left[\alpha\left(\frac{\partial^2 f}{\partial x^2} + \frac{\partial^2 f}{\partial y^2}\right) + \beta\left(x^2+y^2\right) f\right] \\
&+ 2\alpha \frac{\partial \xi}{\partial x}\frac{\partial^2 f}{\partial x^2} + \left(2\alpha \frac{\partial \xi}{\partial y} + 2\alpha \frac{\partial \eta}{\partial x}\right)\frac{\partial^2 f}{\partial x \partial y} + 2\alpha \frac{\partial \eta}{\partial y}\frac{\partial^2 f}{\partial y^2} \\
&+ \left(\alpha \frac{\partial^2 \xi}{\partial x^2} + \alpha \frac{\partial^2 \xi}{\partial y^2} + 2\alpha \frac{\partial \tau}{\partial x}\right)\frac{\partial f}{\partial x} \\
&+ \left(\alpha \frac{\partial^2 \eta}{\partial x^2} + \alpha \frac{\partial^2 \eta}{\partial y^2} + 2\alpha \frac{\partial \tau}{\partial y}\right)\frac{\partial f}{\partial y} \\
&+ \left(\alpha \frac{\partial^2 \tau}{\partial x^2} + \alpha \frac{\partial^2 \tau}{\partial y^2}\right) f
\end{aligned} \tag{6}
$$

$$
\begin{aligned}
\mathbf{XH} f &= \left(\xi \frac{\partial}{\partial x} + \eta \frac{\partial}{\partial y} + \tau\right)\left[\alpha\left(\frac{\partial^2 f}{\partial x^2} + \frac{\partial^2 f}{\partial y^2}\right) + \beta\left(x^2+y^2\right) f\right] \\
&= \alpha\xi \frac{\partial^3 f}{\partial x^3} + \alpha\xi \frac{\partial^3 f}{\partial y^2 \partial x} + \beta\xi\left(x^2 \frac{\partial f}{\partial x} + 2xf + y^2 \frac{\partial f}{\partial x}\right) \\
&\quad + \alpha\eta \frac{\partial^3 f}{\partial x^2 \partial y} + \alpha\eta \frac{\partial^3 f}{\partial y^3} + \beta\eta\left(x^2 \frac{\partial f}{\partial y} + 2yf + y^2 \frac{\partial f}{\partial y}\right) \\
&\quad + \tau\left(\alpha \frac{\partial^2 f}{\partial x^2} + \alpha \frac{\partial^2 f}{\partial y^2} + \beta x^2 f + \beta y^2 f\right) \\
&= \xi\left[\alpha\left(\frac{\partial^3 f}{\partial x^3} + \frac{\partial^3 f}{\partial y^2 \partial x}\right) + \beta\left(\left(x^2+y^2\right)\frac{\partial f}{\partial x} + 2xf\right)\right] \\
&\quad + \eta\left[\alpha\left(\frac{\partial^3 f}{\partial x^2 \partial y} + \frac{\partial^3 f}{\partial y^3}\right) + \beta\left(\left(x^2+y^2\right)\frac{\partial f}{\partial y} + 2yf\right)\right] \\
&\quad + \tau\left[\alpha\left(\frac{\partial^2 f}{\partial x^2} + \frac{\partial^2 f}{\partial y^2}\right) + \beta\left(x^2+y^2\right) f\right]
\end{aligned} \tag{7}
$$

$$
\begin{aligned}
-[\mathbf{X}, \mathbf{H}] f &= \mathbf{HX} f - \mathbf{XH} f \\
&= -2\xi\beta x f - 2\eta\beta y f + 2\alpha \frac{\partial \xi}{\partial x}\frac{\partial^2 f}{\partial x^2} + 2\alpha \frac{\partial \eta}{\partial y}\frac{\partial^2 f}{\partial y^2} \\
&\quad + \left(2\alpha \frac{\partial \xi}{\partial y} + 2\alpha \frac{\partial \eta}{\partial x}\right)\frac{\partial^2 f}{\partial x \partial y}
\end{aligned}
$$

$$
\begin{aligned}
&+\left(\alpha\frac{\partial^2\xi}{\partial x^2}+\alpha\frac{\partial^2\xi}{\partial y^2}+2\alpha\frac{\partial\tau}{\partial x}\right)\frac{\partial f}{\partial x}\\
&+\left(\alpha\frac{\partial^2\eta}{\partial x^2}+\alpha\frac{\partial^2\eta}{\partial y^2}+2\alpha\frac{\partial\tau}{\partial y}\right)\frac{\partial f}{\partial y}\\
&+\left(\alpha\frac{\partial^2\tau}{\partial x^2}+\alpha\frac{\partial^2\tau}{\partial y^2}\right)f\\
&=\left[2\alpha\frac{\partial\xi}{\partial x}\frac{\partial^2}{\partial x^2}+\left(2\alpha\frac{\partial\xi}{\partial y}+2\alpha\frac{\partial\eta}{\partial x}\right)\frac{\partial^2}{\partial x\partial y}+2\alpha\frac{\partial\eta}{\partial y}\frac{\partial^2}{\partial y^2}\right]f\\
&+\left[\left(\alpha\frac{\partial^2\xi}{\partial x^2}+\alpha\frac{\partial^2\xi}{\partial y^2}+2\alpha\frac{\partial\tau}{\partial x}\right)\frac{\partial}{\partial x}\right]f\\
&+\left[\left(\alpha\frac{\partial^2\eta}{\partial x^2}+\alpha\frac{\partial^2\eta}{\partial y^2}+2\alpha\frac{\partial\tau}{\partial y}\right)\frac{\partial}{\partial y}\right]f\\
&+\left(\alpha\frac{\partial^2\tau}{\partial x^2}+\alpha\frac{\partial^2\tau}{\partial y^2}-2\xi\beta x-2\eta\beta y\right)f.
\end{aligned}
$$

If we take $[\mathbf{X},\mathbf{H}]=R\mathbf{H}$, then

$$
-[\mathbf{X},\mathbf{H}]=\left(-R\alpha\frac{\partial^2}{\partial x^2}-R\alpha\frac{\partial^2}{\partial y^2}-R\beta x^2-R\beta y^2\right)f.
$$

This is true for any $f\in\mathcal{V}$. Therefore, the coefficients of $\frac{\partial^2}{\partial x^2}$, $\frac{\partial^2}{\partial y^2}$, $\frac{\partial^2}{\partial x\partial y}$, $\frac{\partial}{\partial x}$ and $\frac{\partial}{\partial y}$ are equal. We thus obtain the system of equations

$$
2\alpha\frac{\partial\xi}{\partial x}=-R\alpha=2\alpha\frac{\partial\eta}{\partial y}\tag{8}
$$

$$
2\alpha\frac{\partial\xi}{\partial y}+2\alpha\frac{\partial\eta}{\partial x}=0\tag{9}
$$

$$
\alpha\frac{\partial^2\xi}{\partial x^2}+\alpha\frac{\partial^2\xi}{\partial y^2}+2\alpha\frac{\partial\tau}{\partial x}=0\tag{10}
$$

$$
\alpha\frac{\partial^2\eta}{\partial x^2}+\alpha\frac{\partial^2\eta}{\partial y^2}+2\alpha\frac{\partial\tau}{\partial y}=0\tag{11}
$$

$$
\alpha\frac{\partial^2\tau}{\partial x^2}+\alpha\frac{\partial^2\tau}{\partial y^2}-2\xi\beta x-2\eta\beta y=-R\beta\left(x^2+y^2\right).\tag{12}
$$

From (8)

$$
\frac{\partial\xi}{\partial x}=\frac{\partial\eta}{\partial y}
$$

$$\Rightarrow \frac{\partial^2 \xi}{\partial x^2} = \frac{\partial^2 \eta}{\partial y \partial x}. \tag{13}$$

From (9)

$$\begin{aligned} \frac{\partial \xi}{\partial y} &= -\frac{\partial \eta}{\partial x} \\ \Rightarrow \frac{\partial^2 \xi}{\partial y^2} &= -\frac{\partial^2 \eta}{\partial x \partial y}. \end{aligned} \tag{14}$$

Adding (13) and (14) we get

$$\frac{\partial^2 \xi}{\partial x^2} + \frac{\partial^2 \xi}{\partial y^2} = \frac{\partial^2 \eta}{\partial y \partial x} - \frac{\partial^2 \eta}{\partial x \partial y} = 0. \tag{15}$$

Similarly, from (8)

$$\frac{\partial^2 \xi}{\partial x \partial y} = \frac{\partial^2 \eta}{\partial y^2}. \tag{16}$$

Also, from (9),

$$\frac{\partial^2 \xi}{\partial y \partial x} = -\frac{\partial^2 \eta}{\partial x^2}. \tag{17}$$

Subtracting (17) and (16),

$$\frac{\partial^2 \eta}{\partial y^2} + \frac{\partial^2 \eta}{\partial x^2} = 0. \tag{18}$$

Substituting (15) and (18) into (10) and (11), respectively, we obtain

$$\frac{\partial \tau}{\partial x} = 0 = \frac{\partial \tau}{\partial y},$$

which yields the solution $\tau = d$ where d is a constant. Thus, (12) becomes

$$-2\beta\,(\xi x + \eta y) = -R\beta\left(x^2 + y^2\right)$$

which gives

$$R = \frac{2\,(\xi x + \eta y)}{\left(x^2 + y^2\right)}. \tag{19}$$

In order to solve the system of nonlinear partial differential equations obtained from (9), we substitute (19) into (8) to obtain

$$\frac{\partial \xi}{\partial x} + \frac{(\xi x + \eta y)}{\left(x^2 + y^2\right)} = 0$$
$$\frac{\partial \eta}{\partial y} + \frac{(\xi x + \eta y)}{\left(x^2 + y^2\right)} = 0$$
$$\frac{\partial \xi}{\partial y} + \frac{\partial \eta}{\partial x} = 0.$$

Using Maple 13 to solve the above system, we get

$$\xi(x, y) = -\left[\frac{cy^3 + \left(cx^2 - a\right) y + bx}{x^2 + y^2}\right]$$

and

$$\eta(x, y) = \frac{cx^3 + \left(a + cy^2\right) x + by}{x^2 + y^2},$$

where $a, b, c \in \mathbb{C}$. Hence the symmetry operator is

$$\begin{aligned}
\mathbf{X} &= -\left[\frac{cy^3 + \left(cx^2 - a\right) y + bx}{x^2 + y^2}\right]\frac{\partial}{\partial x} + \left[\frac{cx^3 + \left(a + cy^2\right) x + by}{x^2 + y^2}\right]\frac{\partial}{\partial y} + d \\
&= \frac{ay}{x^2 + y^2}\frac{\partial}{\partial x} + \frac{ax}{x^2 + y^2}\frac{\partial}{\partial y} - \frac{bx}{x^2 + y^2}\frac{\partial}{\partial x} + \frac{by}{x^2 + y^2}\frac{\partial}{\partial y} \\
&\quad - \frac{cy^3}{x^2 + y^2}\frac{\partial}{\partial x} - \frac{cx^2 y}{x^2 + y^2}\frac{\partial}{\partial x} + \frac{cx^3}{x^2 + y^2}\frac{\partial}{\partial y} - \frac{cy^2 x}{x^2 + y^2}\frac{\partial}{\partial y} + d \\
&= \frac{a}{x^2 + y^2}\left[y\frac{\partial}{\partial x} + x\frac{\partial}{\partial y}\right] + \frac{b}{x^2 + y^2}\left[y\frac{\partial}{\partial y} - x\frac{\partial}{\partial x}\right] \\
&\quad + \frac{cx\left(x^2 + y^2\right)}{x^2 + y^2}\frac{\partial}{\partial y} - \frac{cy\left(x^2 + y^2\right)}{x^2 + y^2}\frac{\partial}{\partial x} + d \\
&= \frac{a}{x^2 + y^2}\left[y\frac{\partial}{\partial x} + x\frac{\partial}{\partial y}\right] + \frac{b}{x^2 + y^2}\left[y\frac{\partial}{\partial y} - x\frac{\partial}{\partial x}\right] \\
&\quad + c\left[x\frac{\partial}{\partial y} - y\frac{\partial}{\partial x}\right] + d,
\end{aligned}$$

and the associated symmetry algebra L is given by

$$L = \left\{\frac{a}{x^2 + y^2}\left(y\frac{\partial}{\partial x} + x\frac{\partial}{\partial y}\right) + \frac{b}{x^2 + y^2}\left(y\frac{\partial}{\partial y} - x\frac{\partial}{\partial x}\right)\right.$$

$$+c\left(x\frac{\partial}{\partial y}-y\frac{\partial}{\partial x}\right)+d\,\middle|\,a,b,c,d\in\mathbb{C}\Big\}.$$

Let

$$A_1=\frac{1}{x^2+y^2}\left(y\frac{\partial}{\partial x}+x\frac{\partial}{\partial y}\right),\ A_2=\frac{1}{x^2+y^2}\left(y\frac{\partial}{\partial y}-x\frac{\partial}{\partial x}\right),\ A_3=x\frac{\partial}{\partial y}-y\frac{\partial}{\partial x}$$

and $B=1$. The set $P=\{A_1,A_2,A_3,B\}$ is a basis for L. We consider the real three-dimensional algebra L^* with basis $\{A_1,A_2,A_3\}$, since $B=1$ is the trivial symmetry operator. The following relations hold:

$$[A_1,A_2]=0,\ [A_2,A_1]=0,\ [A_2,A_3]=-2A_1,\ [A_3,A_2]=2A_1,\ [A_1,A_3]=2A_2,$$

$$[A_3,A_1]=-2A_2,\ [A_1,A_1]=0,\ [A_2,A_2]=0,\ [A_3,A_3]=0.$$

We have not been able to match these structural constants with any known classical Lie algebra. However, the presence of A_3 indicates that Eq. (3) possesses rotational symmetry according to Ibragimov [5]. It is noteworthy that Winterniz [9] showed that the 2D harmonic oscillator exhibits $\mathcal{SU}(2)$ symmetry. Furthermore, Hall [4] showed that the Lie algebra $\mathcal{SU}(2)$ is isomorphic to the Lie algebra $\mathcal{SO}(3)$ and that the Lie group $\mathcal{SU}(2)$ is locally isomorphic to the Lie group $\mathcal{SO}(3)$.

3 Solving the Base Case Equation

The base case equation is given by

$$\frac{\partial^2\psi}{\partial x^2}+\frac{\partial^2\psi}{\partial y^2}+\omega^2\psi=0. \tag{20}$$

Firstly, we reformulate (20) in terms of the complex variables $z=x+iy, \overline{z}=x-iy$ and $\psi=\psi(z,\overline{z})$.

We then introduce the differential operators $\frac{\partial}{\partial z}=\frac{1}{2}\left(\frac{\partial}{\partial x}-i\frac{\partial}{\partial y}\right)$ and $\frac{\partial}{\partial \overline{z}}=\frac{1}{2}\left(\frac{\partial}{\partial x}+i\frac{\partial}{\partial y}\right)$, and note that $\frac{\partial \overline{z}}{\partial z}=0=\frac{\partial z}{\partial \overline{z}}$.

According to Morse [6], we can treat z and $\overline{z}$ as independent variables. Now,

$$\frac{\partial^2\psi}{\partial x^2}+\frac{\partial^2\psi}{\partial y^2}=4\frac{\partial^2\psi}{\partial z\partial\overline{z}}. \tag{21}$$

Substituting (21) into (20),

$$\frac{\partial^2\psi}{\partial z\partial\overline{z}}+\frac{\omega^2}{4}\psi=0. \tag{22}$$

Since we have changed coordinate systems, we provide the following arguments that the same symmetries hold for the variables z and $\overline{z}$. Translations in the z- and $\overline{z}$-directions are given by $z^* = z + a\epsilon$ and $\overline{z}^* = \overline{z} + \overline{a}\epsilon$, where $a, \overline{a} \in \mathbb{C}$ and $\epsilon \in \mathbb{R}$ is a continuous parameter. Since

$$\frac{\partial \psi}{\partial \overline{z}^*} = \frac{\partial \psi}{\partial \overline{z}},$$

it follows that

$$\frac{\partial^2 \psi}{\partial z^* \partial \overline{z}^*} = \frac{\partial^2 \psi}{\partial z \partial \overline{z}}.$$

In terms of z^* and $\overline{z}^*$ Eq. (22) becomes

$$\frac{\partial^2 \psi\left(z^*, \overline{z}^*\right)}{\partial z \partial \overline{z}} + \frac{\omega^2}{4} \psi\left(z^*, \overline{z}^*\right) = 0,$$

where $\psi\left(z^*, \overline{z}^*\right) = \psi\left(z + a\epsilon, \overline{z} + \overline{a}\epsilon\right)$. Rotations about the origin with respect to z and $\overline{z}$ are given by $z' = e^{i\theta} z$, and $\overline{z}' = e^{-i\theta} \overline{z}$, where $\theta \in \mathbb{R}$ is a continuous parameter. Since $\frac{\partial z}{\partial \overline{z}'} = 0$ and $\frac{\partial \overline{z}}{\partial \overline{z}'} = e^{i\theta}$, we have

$$\frac{\partial \psi}{\partial \overline{z}'} = \frac{\partial \psi}{\partial \overline{z}} e^{i\theta} \text{ and } \frac{\partial^2 \psi}{\partial z' \partial \overline{z}'} = \frac{\partial^2 \psi}{\partial z \partial \overline{z}}.$$

Therefore in terms of z' and $\overline{z}'$ Eq. (22) becomes

$$\frac{\partial^2 \psi\left(z', \overline{z}'\right)}{\partial z' \partial \overline{z}'} + \frac{\omega^2}{4} \psi\left(z', \overline{z}'\right) = 0.$$

If $\psi\left(z, \overline{z}\right)$ is a solution to (22) then so is $\psi\left(z', \overline{z}'\right) = \psi\left(e^{i\theta} z, e^{-i\theta} \overline{z}\right)$. Hence both translational and rotational symmetries are preserved after the variable changes to z and $\overline{z}$. According to Ovsiannikov [7], both these symmetries guarantee us the existence of two separate one-parameter continuous transformation groups. We now determine the first group G_1 by making use of the Taylor expansion of $\psi\left(z + a\epsilon, \overline{z} + \overline{a}\epsilon\right)$, parametrized by ϵ about $\epsilon = 0$:

$$\psi\left(z + a\epsilon, \overline{z} + \overline{a}\epsilon\right) = \psi\left(z + a\epsilon, \overline{z} + \overline{a}\epsilon\right)\big|_{\epsilon=0} + \left.\frac{d\psi}{d\epsilon}\right|_{\epsilon=0} + \left.\frac{d^2\psi}{d\epsilon^2}\right|_{\epsilon=0}. \tag{23}$$

Substituting the equations

$$\left.\frac{d\psi}{d\epsilon}\right|_{\epsilon=0} = \left(a \frac{\partial \psi}{\partial z} + \overline{a} \frac{\partial \psi}{\partial \overline{z}}\right) \tag{24}$$

and

$$\left.\frac{d^2\psi}{d\epsilon^2}\right|_{\epsilon=0} = \left(a\frac{\partial}{\partial z} + \overline{a}\frac{\partial}{\partial \overline{z}}\right)^2 \psi \tag{25}$$

into (23),

$$\psi\left(z + a\epsilon, \overline{z} + \overline{a}\epsilon\right) = g\left(\epsilon\right)\psi\left(z, \overline{z}\right), \text{ where } g\left(\epsilon\right) = exp\left[\epsilon\left(a\frac{\partial}{\partial z} + \overline{a}\frac{\partial}{\partial \overline{z}}\right)\right].$$

The group G_1 obtained from the translational symmetry is given by

$$G_1 = \left\{g\left(\epsilon\right)\middle|\, g\left(\epsilon\right) = exp\left[\epsilon\left(a\frac{\partial}{\partial z} + \overline{a}\frac{\partial}{\partial \overline{z}}\right)\right], a \in \mathbb{C}, \epsilon \in \mathbb{R}\right\}.$$

Lie groups are obtained by taking the exponential of the basis vectors of the corresponding Lie algebras. It can easily be verified that G_1 satisfies the conditions for an abelian group: the closure, associativity, and commutativity properties follow directly from the laws of exponents. Furthermore, the identity and inverse elements are given by $g\left(0\right)$ and $g\left(-\epsilon\right)$, respectively.

Similarly, we determine the second group G_2 by expanding $\psi\left(e^{i\theta}z, e^{-i\theta}\overline{z}\right)$ parametrized by θ about $\theta = 0$. Now,

$$\psi\left(e^{i\theta}z, e^{-i\theta}\overline{z}\right) = \psi\left(e^{i\theta}z, e^{-i\theta}\overline{z}\right)\Big|_{\theta=0} + \left(\theta - 0\right)\left(\left.\frac{d\psi}{d\theta}\right|_{\theta=0}\right)$$

$$+ \frac{\left(\theta - 0\right)^2}{2!}\left(\left.\frac{d^2\psi}{d\theta^2}\right|_{\theta=0}\right) + \ldots. \tag{26}$$

Since

$$\begin{aligned}\frac{\partial \overline{z}'}{\partial \theta} &= -ie^{-i\theta}z = -i\overline{z}' \text{ and}\\ \frac{\partial z'}{\partial \theta} &= ie^{i\theta}z \;\; = iz',\end{aligned}$$

it follows that

$$\left.\frac{d\psi}{d\theta}\right|_{\theta=0} = \left(iz\frac{\partial \psi}{\partial z} - i\overline{z}\frac{\partial \psi}{\partial \overline{z}}\right). \tag{27}$$

Consequently,

$$\left.\frac{d^2\psi}{d\theta^2}\right|_{\theta=0} = i\left(z\frac{\partial}{\partial z} - \overline{z}\frac{\partial}{\partial \overline{z}}\right)^2 \psi\left(z, \overline{z}\right). \tag{28}$$

Substituting Eqs. (27) and (28) into (26)

$$\psi\left(e^{i\theta}z, e^{-i\theta}\overline{z}\right) = h(\theta)\,\psi(z,\overline{z}), \text{ where} h(\theta) = exp\left[i\theta\left(z\frac{\partial}{\partial z} - \overline{z}\frac{\partial}{\partial \overline{z}}\right)\right].$$

Hence, the group G_2 obtained from the rotational symmetry is given by

$$G_2 = \left\{h(\theta)|\, h(\theta) = exp\left[i\theta\left(z\frac{\partial}{\partial z} - \overline{z}\frac{\partial}{\partial \overline{z}}\right)\right], \theta \in \mathbb{R}\right\}.$$

We now find a similarity variable $\eta(z,\overline{z})$ with respect to the translational symmetries:

$$z^* = z + a\epsilon \tag{29}$$

$$\overline{z}^* = \overline{z} + \overline{a}\epsilon \tag{30}$$

which, after rearranging, gives

$$\frac{z^*}{a} - \frac{\overline{z}^*}{\overline{a}} = \frac{z}{a} - \frac{\overline{z}}{\overline{a}}.$$

Hence we define the similarity variable

$$\eta(z,\overline{z}) = \frac{z}{a} - \frac{\overline{z}}{\overline{a}}.$$

According to Stephani [8], a similarity variable will remain invariant under the action of the group associated with that variable. Therefore, $\eta(z,\overline{z})$ should remain invariant under the action of the group G_1. We now illustrate this using the following argument: Since

$$\eta\left(z^*, \overline{z}^*\right) = \eta(z,\overline{z})$$

then we have

$$\frac{\partial \psi}{\partial \overline{z}} = \frac{\partial \psi}{\partial \eta}\frac{\partial \eta}{\partial \overline{z}} = \frac{-1}{\overline{a}}\frac{\partial \psi}{\partial \eta}$$

and

$$\frac{\partial^2 \psi}{\partial z \partial \overline{z}} = \frac{-1}{|a|^2}\frac{\partial^2 \psi}{\partial \eta^2}.$$

Since ψ and η only, we can write all partial derivatives as total derivatives. Thus, Eq. (22) becomes

$$\frac{d^2\psi}{d\eta^2} - \left(\frac{|a|\,\omega}{2}\right)^2 \psi = 0.$$

Thus,

$$\psi(\eta) = C_1 e^{(\delta z - \overline{\delta z})} + C_2 e^{-(\delta z - \overline{\delta z})}, \tag{31}$$

where $\delta = \frac{\omega|a|}{2a}$.

Now, consider $\delta z - \overline{\delta z} = \delta z - \overline{\delta z}$. For any complex number $\omega, \omega - \overline{\omega} = 2i\,\text{Im}(\omega)$ which implies that $\delta z - \overline{\delta z} = i\,[2\text{Im}(\delta z)]$. Thus, equation (31) becomes

$$\psi(z, \overline{z}) = D_1 \cos(2\text{Im}(\delta z)) + D_2 \sin(2\text{Im}(\delta z)), \tag{32}$$

where $D_1 = C_1 + C_2$ and $D_2 = i\,(C_1 - C_2)$, Eq. (32) suggests that $\cos(2\text{Im}(\delta z))$ and $\sin(2\text{Im}(\delta z))$ are independent solutions for equation (22). This can be verified via substitution into (22) and by using the facts that $\delta = \frac{\omega|a|}{2a}$ and $\omega = 2|\delta|$. Consider now $a = a_1 + a_2 i \in \mathbb{C}$. Then

$$\delta = \frac{\omega a_1}{2\sqrt{a_1^2 + a_2^2}} - \frac{\omega a_2 i}{2\sqrt{a_1^2 + a_2^2}}.$$

Taking

$$k = \text{Re}(\delta) = \frac{\omega a_1}{2\sqrt{a_1^2 + a_2^2}}, \tag{33}$$

$$l = \text{Im}(\delta) = -\frac{\omega a_2}{2\sqrt{a_1^2 + a_2^2}}, \tag{34}$$

then

$$\delta z = \frac{\omega}{2\sqrt{a_1^2 + a_2^2}} \left[(x a_1 + a_2 y) \quad i\,(a_1 y - a_2 x)\right]$$

Hence,

$$\psi(x, y) = A \cos(2(ky + lx)) + B \sin(2(ky + lx)) \tag{35}$$

is a solution for Eq. (20), where $A, B \in \mathbb{R}$, and l and k are to be determined using any specified boundary conditions for Eq. (20). We now make use of the rotational symmetries to find a second similarity variable $\gamma(z, \overline{z})$. The origin with respect to z and $\overline{z}$ is given by

$$z' = e^{i\theta} z, \tag{36}$$

$$\overline{z}' = e^{-i\theta} \overline{z}. \tag{37}$$

From equation (36) we get

$$\frac{z'}{z} = \frac{\ln\left(\frac{z'}{z}\right)}{i}, \tag{38}$$

and substituting Eq. (38) into (37) yields

$$\overline{z}' = z\overline{z}.$$

If we let $\gamma(z, \overline{z}) = z\overline{z}$, then

$$\gamma(z', \overline{z}') = \gamma(z, \overline{z}).$$

4 Solving the 2D Harmonic Oscillator

To obtain general solutions for the 2D harmonic oscillator, recall the associated equation:

$$\frac{-\hbar^2}{2M}\left(\frac{\partial^2\psi}{\partial x^2} + \frac{\partial^2\psi}{\partial y^2}\right) + \frac{1}{2}M\omega^2\left(x^2 + y^2\right)\psi = E\psi. \tag{39}$$

Let $z = x + iy$, $\overline{z} = x - iy$, $\frac{\partial}{\partial z} = \frac{1}{2}\left(\frac{\partial}{\partial x} - i\frac{\partial}{\partial y}\right)$, and $\frac{\partial}{\partial \overline{z}} = \frac{1}{2}\left(\frac{\partial}{\partial x} + i\frac{\partial}{\partial y}\right)$. We see that z and $\overline{z}$ can be treated as independent variables. Furthermore,

$$\frac{\partial^2\psi}{\partial x^2} + \frac{\partial^2\psi}{\partial y^2} = 4\frac{\partial^2\psi}{\partial z\partial\overline{z}}. \tag{40}$$

Consider now

$$z + \overline{z} = \frac{z^2 + \overline{z}^2 + 2z\overline{z}}{4}$$

and

$$z - \overline{z} = \frac{-\left(z^2 + \overline{z}^2 - 2z\overline{z}\right)}{4}.$$

After substituting

$$x^2 + y^2 = z\overline{z} \tag{41}$$

and (40) in (39), we obtain

$$\Rightarrow \frac{\partial^2\psi}{\partial z\partial\overline{z}} - \alpha z\overline{z}\psi + \frac{k^2}{4}\psi = 0, \tag{42}$$

where $\alpha = \frac{M\omega^2}{2\hbar}$, $k = \sqrt{\frac{2ME}{\hbar^2}}$, and $\psi = \psi(z, \overline{z})$.

The origin with respect to z and $\overline{z}$ is given by $z' = e^{i\theta}z$, and $\overline{z}' = e^{-i\theta}\overline{z}$, where $\theta \in \mathbb{R}$ is a continuous parameter and

$$\frac{\partial^2\psi}{\partial z'\partial\overline{z}'} = \frac{\partial^2\psi}{\partial z\partial\overline{z}}.$$

Furthermore,

$$z'\overline{z}' = \left(e^{i\theta}z\right)\left(e^{-i\theta}\overline{z}\right) = z\overline{z}$$

and Eq. (42) becomes

$$\frac{\partial^2\psi\left(z'\cdot\overline{z}'\right)}{\partial z'\partial\overline{z}'} - \alpha z'\overline{z}'\psi\left(z'\cdot\overline{z}'\right) + \frac{k^2}{4}\psi\left(z'\cdot\overline{z}'\right) = 0.$$

Hence if $\psi\left(z,\overline{z}\right)$ is a solution to (42) then so is $\psi\left(z'\cdot\overline{z}'\right) = \psi\left(e^{i\theta}z, e^{-i\theta}\overline{z}\right)$. Thus, rotational symmetry is preserved even after changing to the complex variables z and $\overline{z}$.

We know that a group G_2 is obtained from the rotation symmetry as given by

$$G_2 = \left\{h\left(\theta\right)\left|h\left(\theta\right) = exp\left[i\theta\left(z\frac{\partial}{\partial z} - \overline{z}\frac{\partial}{\partial\overline{z}}\right)\right], \theta \in \mathbb{R}\right.\right\},$$

and $\xi\left(z,\overline{z}\right) = \left(z\overline{z}\right)^m$ where $m \in \mathbb{R}$ is a similarity variable associated with the group G_2. Therefore (42) becomes

$$m^2\xi^{\frac{2m-1}{m}}\frac{\partial^2\psi\left(\xi\right)}{\partial\zeta^?} - \alpha\xi^{\frac{1}{m}}\psi\left(\xi\right) + \frac{k^2}{4}\psi\left(\xi\right) = 0,$$

which has general solution

$$\begin{aligned}\psi\left(\xi\right) &= A\left(\xi\right)N\left(\frac{m+1}{2} - \frac{\omega^2}{8\sqrt{\alpha}}, m+1, 2\sqrt{\alpha}\xi^{\frac{1}{m}}\right)\\ &+ B\left(\xi\right)U\left(\frac{4(m+1)\sqrt{\alpha} - \omega^2}{8\sqrt{\alpha}}, m+1, 2\sqrt{\alpha}\xi^{\frac{1}{m}}\right)\end{aligned} \tag{43}$$

where

$$A\left(\xi\right) = C\xi^{\frac{m-1}{2m}}\left(2\sqrt{\alpha}\xi^{\frac{1}{m}}\right)^{\frac{m+1}{2}}\exp\left(-\sqrt{\alpha}\xi^{\frac{1}{m}}\right),$$

$$B\left(\xi\right) = D\xi^{\frac{m-1}{2m}}\left(2\sqrt{\alpha}\xi^{\frac{1}{m}}\right)^{\frac{2}{4m+1}}\exp\left(-\sqrt{\alpha}\xi^{\frac{1}{m}}\right),$$

$C, D \in \mathbb{R}$, $\alpha = \left(\frac{M\omega}{2\hbar}\right)^2$, $\xi = \left(x^2 + y^2\right)^m$, and N and U denote Kummer's functions of the first kind and second kind, respectively.

5 Conclusion

The symmetry group of the base case equation (recall that this is the two-dimensional Helmholtz equation) was found to be the Euclidean group $E(2)$ which indicated the presence of both rotational and translational symmetries. This result is consistent with the results obtained by who showed that the three-dimensional Helmholtz equation has symmetry group $E(3)$.

The 2D harmonic oscillator was also shown to exhibit rotational symmetry which is consistent with results obtained by Boyer et al. [2]. Doll [3] suggests that rotational invariance implies conservation of angular momentum. This is consistent with Noether's First Theorem which tells us that the presence of any differentiable dynamical symmetry (continuous rotations, translations, etc.) of a physical system is related to a conservation law. More importantly, a basis for the Lie algebra of the 2D harmonic oscillator was found. This basis does not seem to correspond to the basis of any known classical Lie algebras. The commutation relations are consistent and satisfy all the conditions of a Lie algebra. We have not yet been able to correspond our structural constants with the structural coefficients of any three-dimensional real Lie algebra. This suggests the possibility that a new basis for the Lie algebra of the 2D harmonic oscillator has been found. Further investigation into this possibility will be required.

References

1. Bluman, G.W., Anco, S.C.: Symmetry and Integration Methods for Differential Equations. Springer (2002)
2. Boyer, C.P., Kalnins, E.G., Miller, W.: Symmetry and separation of variables for the Helmoltz and Laplace equations. Nagoya Math. J. **60**, 35–80 (1976)
3. Doll, R., Ingold, G.L.: Lissajous curves and semiclassical theory: the two-dimensional harmonic oscillator. Am. J. Phys. **75**(3) (2007)
4. Hall, B.: Lie Groups, Lie Algebras, and Representations: An Elementary Introduction, vol. 222. Springer (2003)
5. Ibragimov, N.K.: Group analysis of ordinary differential equations and the invariance principle in mathematical physics (for the 150th anniversary of sophus lie). Russ. Math. Surv. **47**(4), 89–156 (1992)
6. Morse, P.M., Feshbach, H.: Methods of Theoretical Physics, vol. 1. McGraw-Hill (1953)
7. Ovsiannikov, L.: Group Analysis of Differential Equations. Academic Press (1982)
8. Stephani, H.: Differential Equations: Their Solution Using Symmetries. Cambridge University Press (1989)
9. Winterniz, P., Smorodinskii, Y.A., Uhlir, M., Fris, I.: Symmetry groups in classical and quantum mechanics. J. Nucl. Phys. (U.S.S.R.) **4**, 625–635 (1966)

Analysis and Simulation of Fractional Nonlinear Convection–Diffusion Model with Caputo Derivative

Lalit Mohan and Amit Prakash

Abstract In this research, we investigate the fractional nonlinear convection–diffusion model with the Caputo derivative. The homotopy perturbation transform technique is used to simulate the fractional convection–diffusion model. The error analysis of the presented method is also provided. Lastly, the efficiency of the presented technique is demonstrated by simulating two examples, the results are given in tables and graphs.

Keywords Fractional nonlinear convection–diffusion model (FNCDM) · Caputo derivative · Homotopy perturbation transform technique (HPTT) · Laplace transform · Error analysis

1 Introduction

In recent years, the theory of fractional calculus (FC) has been used to model many intricate physical and natural processes as fractional differential equations (FDE's). Fractional models are more suitable for describing the memory and heredity aspects that result from the materials employed in phenomena than ordinary calculus. The Caputo derivative [1] is among the finest fractional derivatives that we have access to. Analysing how different elements and factors affect the physical process is made easier with the use of mathematical modelling. Numerous academics investigated and analysed a variety of fractional mathematical models, such as the fractional advection–diffusion equation [2], the fractional Korteweg-de Vries equation [3], the fractional integro-partial differential equations [4], the respiratory syncytial virus infection model [5], the problem of heat conduction [6], the fraction BBM-Burger

L. Mohan · A. Prakash (✉)
Department of Mathematics, National Institute of Technology, Kurukshetra, India
e-mail: amitmath@nitkkr.ac.in

L. Mohan
e-mail: lalit_62000037@nitkkr.ac.in

R. K. Sharma et al. (eds.), *Mathematical Theory and Simulation of Scientific Problems*, Springer Proceedings in Mathematics & Statistics 487,
https://doi.org/10.1007/978-981-96-2579-6_8

equation [7], the problem of oil spill [8], the fractional Burger-Huxley equation [9–11], the fractional Burger-Fisher model [12] and brain tumor model with arbitrary order [13].

The convection–diffusion models are frequently used in the modelling of numerous physical processes, including mass and energy movement, generation of the global climate, and dispersion of chemicals in reactors. There has been an increase in interest in the convection–diffusion model in the field of fractional calculus [14–18] since convection–diffusion phenomena can be observed in various natural and physical circumstances, such as heat, mass, and other transport processes. To simulate anomalous diffusion processes, one can use the FNCDM.

The development of effective numerical algorithms is one of the main challenges with the FNCDM, and computational efficiency is a significant parameter for numerical techniques. Numerous numerical techniques, including the mesh-less techniques [19], the spectral technique [20], the finite element technique [21], the finite difference technique [22], the HPTM [23] and the radial basis function method [24], have been proposed for simulating FDE's.

In this article, we assume the following FNCDM

$$ {}_0^cD_t^\alpha u(x,t) = \Delta u(x,t) - \nabla u(x,t) + \Phi(u) + f(x,t), \alpha\varepsilon(0,1] \tag{1} $$

with $u(x, 0) = h(x)$,

where $\Phi(u)$ is non-linear function of u, $u(x, t)$ is the temperature in heat transfer, $f(x, t)$ is source term, $h(x)$ is the smooth function of x, and ${}_0{}^cD_t^\alpha$ is the Caputo operator of order α with $0 < \alpha \leq 1$.

This paper's primary goal is to analyse the FNCDM using the Caputo operator. Analysis is done on the suggested technique's greatest error. The FCDM is solved using the HPTT. The Laplace transform and the Homotopy perturbation method are flawlessly combined in the HPTT. The main contribution of this article is to give an efficient technique for obtaining the solution of the FNCDM. The findings in this article could be extremely helpful for research on oil reservoir models, mass and energy movement, the formation of the world's weather, and chemical dispersion in reactors. To show the efficiency of the suggested technique, we also simulate two test cases.

2 Preliminaries

Definition 2.1 [1] The Caputo derivative $f(t)$ is defined as:

$$ \mathrm{D}^\alpha f(t) = I^{m-\alpha} D^m f(t) = \frac{1}{\Gamma(m-\alpha)} \int_0^t (t-s)^{m-\alpha-1} f^m(s) ds, $$

where $m - 1 < \alpha \leq m$.

Definition 2.2 [1] The Laplace transform (LT) of $y(t)$ is defined as:

$$L[y(t)] = y(s) = \int_0^\infty e^{-st} y(t) dt.$$

Definition 2.3 [1] The LT of the Caputo derivative is given by:

$$L[D_t^\alpha y(t)] = s^\alpha y(s) - \sum_{k=0}^{m-1} s^{(\alpha-k-1)} y^{(k)}(0),\ m-1 < \alpha \le m.$$

3 Description of Proposed Technique

Let us consider the FNCDM as given below

$${}_0{}^c D_t^\alpha u(x,t) = \Delta u(x,t) - \nabla u(x,t) + \Phi(u) + f(x,t), \tag{2}$$

with $u(x, 0) = h(x)$.

Where ${}_0{}^c D_t^\alpha$ is the Caputo operator, $f(x, t)$ is source term and $\Phi(u)$ is nonlinear operator.

Applying the LT to Eq. (2), we have

$$LT\left[{}_0{}^c D_t^\alpha u(x,t)\right] = LT\left[\Delta u(x,t) - \nabla u(x,t) + \Phi(u) + f(x,t)\right]. \tag{3}$$

Using definition (2.3) and initial condition, we get

$$s^\alpha LT[u(x,t)] - s^{\alpha-1} u(x,0) = LT\left[\Delta u(x,t) - \nabla u(x,t) + \Phi(u) + f(x,t)\right],$$

$$LT[u(x,t)] = \frac{1}{s} h(x) + \left(\frac{1}{s}\right)^\alpha LT\left[\Delta u(x,t) - \nabla u(x,t) + \Phi(u) + f(x,t)\right]. \tag{4}$$

Taking the inverse LT on Eq. (4), we have

$$u(x,t) = h(x) + LT^{-1}\left\{\left(\frac{1}{s}\right)^\alpha LT\left[\Delta u(x,t) - \nabla u(x,t) + \Phi(u) + f(x,t)\right]\right\}. \tag{5}$$

Now, we decompose the linear term as

$$L[u(x,t)] = \sum_{l=0}^\infty p^l L[u_l(x,t)].$$

Here $L[u(x,t)] = \left[\Delta u(x,t) - \nabla u(x,t) + f(x,t)\right]$.

The nonlinear term $\Phi(u)$ is decomposed with help of homotopy's polynomial as

$$\Phi(u) = \sum_{l=0}^{\infty} p^l H_l(u),$$

and

$$H_l(u) = \frac{1}{l!}\frac{d^l}{dp^l}\left[\Phi\left(\sum_{j=0}^{l} p^j u_j(x,t)\right)\right]_{p=0}. \tag{6}$$

Using all these equations in (5) we have

$$\sum_{l=0}^{\infty} p^l u_l(x,t) = h(x) + LT^{-1}\left\{\left(\frac{1}{s}\right)^{\alpha} LT\left[\sum_{l=0}^{\infty} p^l L[u_l(x,t)] + \sum_{l=0}^{\infty} p^l H_l(u))\right]\right\}. \tag{7}$$

Comparing the coefficient of like power of p, we get

$$u_0(x,t) = h(x) + LT^{-1}\left\{\left(\frac{1}{s}\right)^{\alpha} LT[f(x,t])\right\},$$

$$u_1(x,t) = LT^{-1}\left\{\left(\frac{1}{s}\right)^{\alpha} LT[L[u_0(x,t)] + H_0[u]]\right\},$$

$$u_2(x,t) = LT^{-1}\left\{\left(\frac{1}{s}\right)^{\alpha} LT[L[u_1(x,t)] + H_1[u]]\right\},$$

$$\vdots$$

$$u_l(x,t) = LT^{-1}\left\{\left(\frac{v}{s}\right)^{\alpha} LT\left[L\left[u_{l-1}(x,t)\right] + H_{l-1}[u]\right]\right\}.$$

The HPTT solution of (2) is given by

$$u(x,t) = \lim_{k\to\infty} \sum_{l=0}^{k} u_l(x,t). \tag{8}$$

4 Error Analysis of HPTT

In this section, we approximate the error of the presented method.

Theorem 4.1 *If* $\exists$ *a constant* $0 < \varepsilon < 1$ *such that* $\|u_{n+1}(x,t)\| \leq \varepsilon \|u_n(x,t)\|\ \forall n$ *and if the series* $\sum_{n=0}^{r} u_n(x,t)$ *is treated as numerical solution* $u(x,t)$*, then the absolute error is given by.*

$$\|u(x,t) - \sum_{n=0}^{r} u_n(x,t)\| \leq \frac{\varepsilon^{r+1}}{(1-\varepsilon)} \|u_0(x,t)\|$$

Proof We have.

$$\left\| u(x,t) - \sum_{n=0}^{r} u_n(x,t) \right\| = \left\| \sum_{n=r+1}^{\infty} u_n(x,t) \right\|,$$
$$\leq \sum_{n=r+1}^{\infty} \|u_n(x,t)\|,$$
$$\leq \sum_{n=r+1}^{\infty} \varepsilon^m \|u_0(x,t)\|,$$
$$\leq (\varepsilon)^{r+1}\left[1 + (\varepsilon)^1 + (\varepsilon)^2 + \ldots\right] \|u_0(x,t)\|,$$
$$\leq \tfrac{\varepsilon^{r+1}}{(1-\varepsilon)} \|u_0(x,t)\|$$

5 Application and Results Discussion

In this segment, two examples of the FNCDM are solved by using the presented method, to illustrate the accuracy of presented method, HPTT.

Example 5.1 Let us assume the following FNCDM.

$${}_0^{c}D_t^{\alpha} u(x,t) = \Delta u(x,t) - \nabla u(x,t) + u(x,t).u_{xx}(x,t) - u^2(x,t) + u(x,t), \quad (9)$$

with $u(x,0) = e^x, 0 < \alpha \leq 1$ and the analytical solution of Eq. (9) at $\alpha = 1$ is $u(x,t) = e^{x+t}$.

Solution. Applying the HPTT to Eq. (9), we obtain

$$u_0(x,t) = e^x,$$

$$u_1(x,t) = e^x \frac{t^{\alpha}}{\Gamma(\alpha+1)},$$

$$u_2(x,t) = e^x \frac{t^{2\alpha}}{\Gamma(2\alpha+1)},$$

$$u_3(x,t) = e^x \frac{t^{3\alpha}}{\Gamma(3\alpha+1)},$$

$$\vdots$$

and

$$u(x,t) = u_0(x,t) + u_1(x,t) + u_2(x,t) + u_3(x,t)$$

Example 5.2 Let us assume the following FNCDM.

$$ {}_0^c D_t^\alpha u(x,t) = \Delta u(x,t) - \nabla u(x,t) + f(x,t), \tag{10}$$

with $f(x,t) = \frac{\Gamma(4+\alpha)t^3}{6} sin\pi x + \pi^2 t^{3+\alpha} sin\pi x + \pi^3 t^{3+\alpha} cos\pi x$, $u(x,0) = 0$, $0 < \alpha \leq 1$ and the analytical solution of equation (10) at $\alpha = 1$ is $u(x,t) = t^{3+\alpha} sin\pi x$.

Solution. Applying the HPTT to equation (5.2), we obtain

$$u_0(x,t) = \Gamma(4+\alpha)\left[\frac{t^{\alpha+3}}{\Gamma(4+\alpha)} sin\pi x + \pi^2 \frac{t^{2\alpha+3}}{\Gamma(4+2\alpha)} sin\pi x + \pi \frac{t^{2\alpha+3}}{\Gamma(4+2\alpha)} cos\pi x\right],$$

$$u_1(x,t) = \Gamma(4+\alpha)\left[\frac{t^{2\alpha+3}}{\Gamma(4+2\alpha)}(-\pi^2 sin\pi x - \pi\, cos\pi x) + \frac{t^{3\alpha+3}}{\Gamma(4+3\alpha)}(-\pi^4 sin\pi x - 2\pi^3 cos\pi x + \pi^2 sin\pi x)\right],$$

$$u_2(x,t) = \Gamma(4+\alpha)\left[\frac{t^{3\alpha+3}}{\Gamma(4+3\alpha)}(\pi^4 sin\pi x + \pi^3 cos\pi x + \pi^2 cos\pi x - \pi^2 sin\pi x) + (\pi^6 sin\pi x + 2\pi^5 cos\pi x - \pi^4 sin\pi x + \pi^5 sin\pi x - 2\pi^4 cos\pi x - \pi^2 sin\pi x) \times \frac{t^{4\alpha+3}}{\Gamma(4+4\alpha)}\right],$$

$$\vdots$$

and

$$u(x,t) = u_0(x,t) + u_1(x,t) + u_2(x,t) + \ldots$$

The absolute errors of HPTT at different values of x, t and α, given in Tables 1 and 2 for Examples 5.1 and 5.2, correspondingly. The comparison of analytical and HPTT solution at $\alpha = 1$, is represented through graphs in Figs. 1, 2, 3 and 4, for Examples 5.1 and 5.2 correspondingly, and it is observed that the HPTT and analytical solution shows exactly same behaviour. The absolute error of HPTT at different values of x, t and α is given in Tables 1 and 2, for Examples 5.1 and 5.2, correspondingly. It can be clearly observed that the absolute error is very less, which shows the efficiency of the presented method.

Table 1 The absolute error of HPTT at $\alpha = 1$, for Example 5.1

x	t	Absolute error
1	1	$1.7\,e^{-15}$
2	2	0
3	3	$5.6\,e^{-14}$
4	4	$4.5\,e^{-13}$
5	5	0
6	6	$2.9\,e^{-11}$
7	7	$2.3\,e^{-10}$
8	8	$7.4\,e^{-09}$
9	9	$7.4\,e^{-09}$
10	10	$1.1\,e^{-07}$

Table 2 The absolute error of the HPTT at different value of α, for Example 5.2

x	t	Absolute error	
		$\alpha = 0.95.$	$\alpha = 1.$
1	0.01	$2.2\,e^{-10}$	$1.2\,e^{-10}$
2	0.02	$6.6\,e^{-09}$	$4.0\,e^{-09}$
3	0.03	$4.8\,e^{-08}$	$3.1\,e^{-08}$
4	0.04	$2.0\,e^{-07}$	$1.3\,e^{-08}$
5	0.05	$6.0\,e^{-07}$	$4.0\,e^{-07}$
6	0.06	$1.4\,e^{-06}$	$1.0\,e^{-06}$
7	0.07	$3.2\,e^{-06}$	$2.2\,e^{-06}$
8	0.08	$6.2\,e^{-06}$	$4.3\,e^{-06}$
9	0.09	$1.1\,e^{-05}$	$7.8\,e^{-06}$
10	0.10	$1.8\,e^{-05}$	$1.3\,e^{-05}$

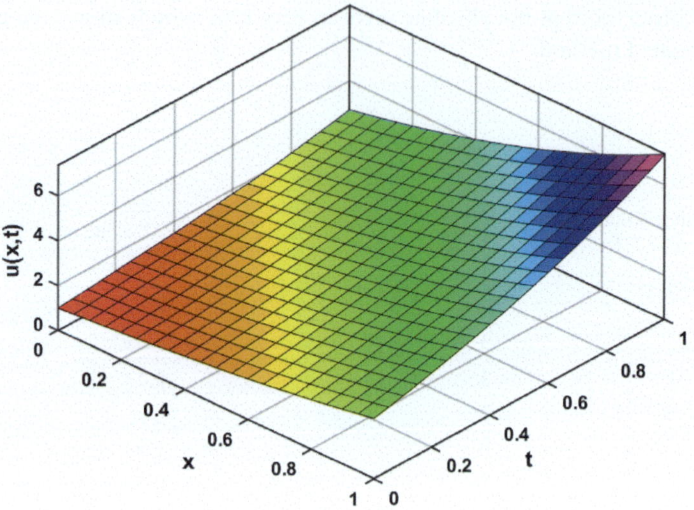

Fig. 1 Exact solution at $\alpha = 1$, for Example 5.1

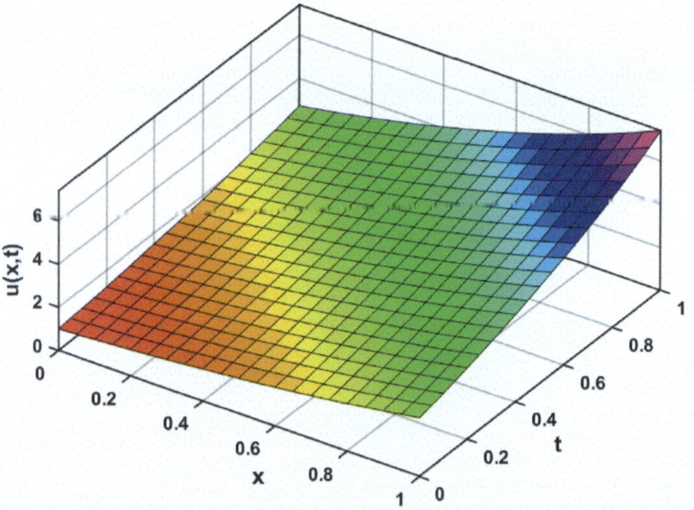

Fig. 2 HPTT solution at $\alpha = 1$, for Example 5.1

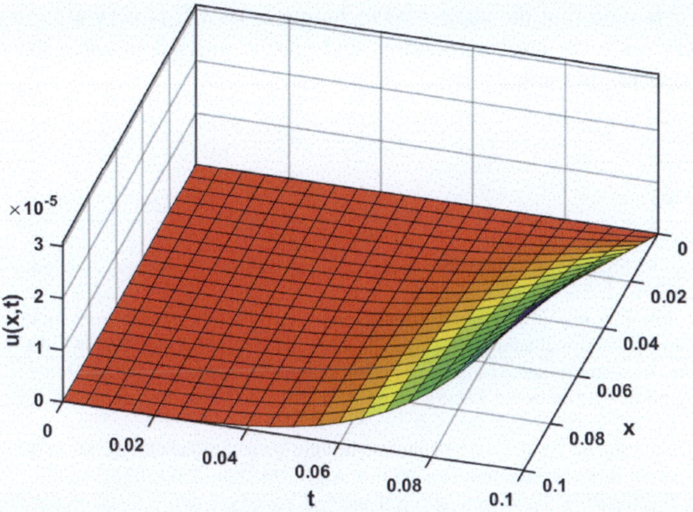

Fig. 3 Exact solution at $\alpha = 1$, for Example 5.2

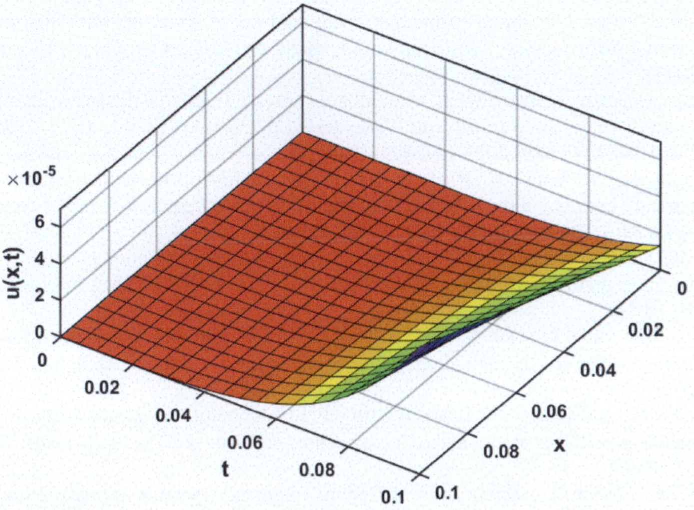

Fig. 4 HPTT solution at $\alpha = 1$, for Example 5.2

6 Conclusion

In this article, we investigate the FNCDM via HPTT with Caputo operator. The error analysis of the presented method is also provided. Two examples of the FNCDM are solved, and the numerical results and graphs show that the HPTT provides an accurate solution that converges to the exact solution with a very small absolute error. As a result, we can state that the suggested techniques, HPTT, is very effective, handle non-linearity and restrictions smoothly, and can be utilised to solve the nonlinear models of real life phenomena.

References

1. Podlubny, I.: Fractional Differential Equations, pp. 1–366. New York, Academic Press, San Diego (1999)
2. Shen, S., Liu, F., Anh, V.: Numerical approximations and solution techniques for the space-time Riesz-Caputo fractional advection-diffusion equation. Numer Algor. **56**, 383–403 (2011)
3. Taghipour, M., Aminikhah, H.: Numerical solution of distributed-order fractional Korteweg-de Vries equation via fractional Zigzag rising diagonal functions. Numer Algor. (2023). https://doi.org/10.1007/s11075-023-01664-0
4. Maji, S., Natesan, S.: Analytical and numerical solution techniques for a class of time-fractional integro-partial differential equations. Numer Algor. **94**, 229–256 (2023)
5. Nemati, S., Torres, D.F.: A new spectral method based on two classes of hat functions for solving systems of fractional differential equations and an application to respiratory syncytial virus infection. Soft. Comput. **25**(9), 6745–6757 (2021)
6. Mohan, L., Prakash, A.: Stability and numerical analysis of the generalised time-fractional Cattaneo model for heat conduction in porous media. Eur. Phys. J. Plus **138**, 1–29 (2023)
7. Mohan, L., Prakash, A.: Stability and numerical analysis of fractional BBM-burger equation and fractional diffusion-wave equation with Caputo derivative. Opt. Quant. Electron. **56**(1), 1–25 (2023)
8. Mohan, L., Prakash, A.: An efficient technique for solving fractional diffusion equation arising in oil pollution via Natural transform. Waves Random Complex Media. 1–25 (2023). https://doi.org/10.1080/17455030.2023.227332
9. Ravi Kanth, A.S.V., Aruna, K., Raghavendar, K: Natural transform decomposition method for the numerical treatment of the time fractional burgers–Huxley equation. Numer. Methods Part. Differ. Equ. **39**(3), 2690–2718 (2023)
10. Idowu, K.O., Loyinmi, A.C.: The analytic solution of non-linear burgers–Huxley equations using the Tanh method. Al-Bahir J. Eng. Pure Sci. **3**(1), 8 (2023). https://doi.org/10.55810/2313-0083.1038
11. Singh, A., Diwedi, K.D., Das, S., Ong, S.H.: Study of one-dimensional space-time fractional-order burgers-fisher and burgers-Huxley fluid models. Math. Methods Appl. Sci. **44**(3), 2455–2467 (2021)
12. Mesgarani, H., Aghdam, Y.E., Jafari, B.: An efficient technique to approximate the nonlinear fractional burgers–fisher model in the Caputo sense. AIP Adv. **13**(1) (2023). https://doi.org/10.1063/5.0130602
13. Prakash, A., Mohan, L.: Application of Caputo fractional operator to analyse the fractional model of brain Tumour via modified technique. Int. J. Appl. Comput. Math. **9**(5), 1–33 (2023)
14. Momani, S., Odibat, Z.M.: Fractional green function for linear time-fractional inhomogeneous partial differential equations in fluid mechanics. J. Appl. Math. Comput. **24**, 167–178 (2007)
15. Saadatmandi, A., Dehghan, M.A.: New operational matrix for solving fractional-order differential equations. Comput. Math. Appl. **59**(3), 1326−1336 (2010)
16. Parvizi, M., Eslahchi, M.R., Dehghan, M.: Numerical solution of fractional advection-diffusion equation with a nonlinear source term. Numer. Algorithms. **68**, 601−629 (2015)
17. Li, C., Zhao, Z., Chen, Y.: Numerical approximation of nonlinear fractional differential equations with subdiffusion and superdiffusion. Comput. Math. Appl. **62**(3), 855–875 (2011)
18. Zhuang, P., Liu, F., Anh, V., Turner, I.: Numerical methods for the variable-order fractional advection-diffusion equation with a nonlinear source term. SIAM J. Numer. Anal. **47**(3), 1760–1781 (2009)

19. Jiang, Y., Ma, J.: High-order finite element methods for time-fractional partial differential equations. J. Comput. Appl. Math. **235**(11), 3285–3290 (2011)
20. Liu, F., Zhuang, P., Anh, V., Turner, I., Burrage, K.: Stability and convergence of the difference methods for the space–time fractional advection–diffusion equation. Appl. Math. Comput. **191**(1), 12–20 (2007)
21. Lin, Y., Xu, C.: Finite difference/spectral approximations for the time-fractional diffusion equation. J. Comput. Phys. **225**(2), 1533–1552 (2007)
22. Rostamian, M., Shahrezaee, A.: A meshless method to the numerical solution of an inverse reaction–diffusion–convection problem. Int. J. Comput. Math. **94**(3), 597–619 (2017)
23. Mohan, L., Prakash, A.: Analysing the conduction of heat in porous medium via Caputo fractional operator with Sumudu transform. J. Comput. Anal. Appl. **33**(1), 1–20 (2024)
24. Liu, Q.X., Gu, Y., Zhuang, P., Liu, F., Nie, Y.: An implicit RBF meshless approach for time fractional diffusion equations. Comput. Mech. **48**, 1–12 (2011)

Common Fixed Point Theorem for Multivalued Mappings in Weak-Partial *b*-Metric Space with an Application

Lucas Wangwe and Santosh Kumar

Abstract This paper proves a common fixed point theorem for multivalued mappings in weak-partial b-metric spaces. We illustrate the operation of our findings with an example. An application of the findings demonstrated here will be the solution of a nonlinear fractional differential equation.

Keywords Common fixed point · Multivalued mapping · Weak-partial b-metric space · Nonlinear fractional differential equation

1 Introduction

The length between two sets of a metric space can be found using the Hausdorff metric. Using concepts from Hausdorff metric, Nadler [24] extended Banach's contraction principle in 1969. Nadler specifically demonstrated that a multivalued mapping has a fixed point if it meets a specific contraction requirement. Numerous fields of mathematics and science, such as nonlinear analysis, optimization, control theory, and dynamical systems, can benefit from this discovery.

Originally introduced by Bakhtin [4] and Czerwik [9], the b-metric spaces provided a generalized notion of metric spaces. Eventually, the concept of multivalued contraction mappings was extended to b-metric spaces by Czerwik [10]. Boriceanu et al. [7], Chifu et al. [8], Kirk and Shahzad [19], Kutbi et al. [21], Roshan et al. [31], and their references are a few of the works in this field.

L. Wangwe (✉)
Department of Mathematics and Statistics, Mbeya University of Science and Technology, Mbeya, Tanzania
e-mail: wangwelucas@gmail.com

S. Kumar
Department of Mathematics, School of Physical Sciences, North-Eastern Hill University, Shillong, Meghalaya, India
e-mail: drsengar2002@gmail.com; santoshkumar@nehu.ac.in

R. K. Sharma et al. (eds.), *Mathematical Theory and Simulation of Scientific Problems*, Springer Proceedings in Mathematics & Statistics 487,
https://doi.org/10.1007/978-981-96-2579-6_9

Matthews [22] provided the first description of non-zero self-distance. It is primarily utilized in programming languages, computer networks, and data structures. The idea introduced by Matthews was expanded by Aydi et al. [3] from single-valued to multivalued mapping in partial metric space. The key ideas about partial metric spaces for single- and multivalued mappings are found in the references [25–27, 33].

Shukla [34] combined the concepts of b-metric space and partial metric space, and utilized these to prove a number of fixed point theorems. Heckmann [14] also provided a weak-partial metric space generalization of partial metric space. An extended version of the multivalued mappings on weak-partial metric spaces was introduced by Beg and Pathak [6]. According to Kanwal et al. [17], the phrases weak-partial b-metric space and weak-partial Hausdorff b-metric space were defined. We direct the reader to [37, 38] for latest research.

The concept of compatible mappings was developed by Kaneko and Sessa [16] after being first put out by Jungck [15]. In accordance with Kubiak and Nadler's work, they proved coincidence fixed point theorems for hybrid pairings of compatible mappings [20, 24]. The notions of weakly compatible Hausdorff mappings in metric spaces were introduced by Pathak [28].

In this work, a shared fixed point theorem for two multivalued mappings in weak-partial b-metric space is established. We consider the space to be paracompact and generalize the results of Kanawal et al. [17], Khojasteh et al. [18], Demma and Vetro [11], and Rhoades [29] Aliouche and Hamaizia [2] from b-metric space to weak-partial b-metric space.

2 Preliminaries

This section presents definitions, lemma, proposition, and some preliminary results. These elements will be utilized to construct the paper's primary findings.

Kanwal et al. [17] gave the notions of a weak-partial b-metric as follows:

Definition 1 ([17]) A distance function $\varrho_b : X \times X \to \mathbf{R}^+$ is known as a weak-partial b-metric on X with $X \neq \emptyset$ and $s \geq 1$ if for all $\theta, \xi, \jmath \in X$, conditions below hold:

(WPB1) $\varrho_b(\theta, \theta) = \varrho_b(\theta, \xi) \Longleftrightarrow \theta = \xi$,
(WPB2) $\varrho_b(\theta, \theta) \leq \varrho_b(\theta, \xi)$,
(WPB3) $\varrho_b(\theta, \xi) = \varrho_b(\xi, \theta)$, and
(WPB4) $\varrho_b(\theta, \xi) \leq s[\varrho_b(\theta, \jmath) + \varrho_b(\jmath, \xi)]$.

It follows that the pair $(\mathcal{X}, \varrho_b)$ is defined as a weak-partial b-metric space.

Example 1 Let $X = \{\frac{1}{2}, \frac{1}{3}, \frac{1}{4}, \frac{1}{5}\}$. Define the generalized ϱ_b on X as

$$\varrho_b\Big(\frac{1}{2}, \frac{1}{4}\Big) \leq \frac{s}{4},$$

$$\varrho_b\left(\frac{1}{3}, \frac{1}{5}\right) \leq \frac{2s}{15}.$$

For $d(\theta, \xi) = d(\xi, \theta)$, for all $\theta, \xi \in X$ and $d(\theta, \xi) = |\theta - \xi|$, we have

$$d\left(\frac{1}{2}, \frac{1}{4}\right) = \frac{1}{4},$$
$$d\left(\frac{1}{3}, \frac{1}{5}\right) = \frac{2}{15}.$$

This shows that $(X, \varrho_b))$ is a weak-partial b-metric space. Since it does not satisfy triangular inequality in $X = \{\frac{1}{2}, \frac{1}{3}, \frac{1}{4}, \frac{1}{5}\}$. It is wider than usual metric space.

An example of weak-partial b-metric space is as follows:

Example 2 ([17])

(1) A distance function $\varrho_b : X \times X \to X$ with (X, ϱ_b) is defined as

$$\varrho_b(\theta, \xi) = |\theta - \xi|^2 + 1,$$

for all $\theta, \xi \in X$.

(2) A distance function $\varrho_b : X \times X \to X$ with (X, ϱ_b) is defined as

$$\varrho_b(\theta, \xi) = \frac{1}{2}|\theta - \xi|^2 + \max\{\theta, \xi\},$$

for all $\theta, \xi \in X$.

Definition 2 ([17]) Assume that the weak-partial b-metric space is (X, ϱ_b). $\{\theta_n\}$ is a series that converges at $\theta \in X$ if and only if

$$\varrho_b(\theta, \theta) = \lim_{n \to \infty} \varrho_b(\theta, \theta_n).$$

Kanwal et al. [17] extended the below definition of convergence.

Definition 3 ([17]) Assume that the weak-partial b-metric space is (X, ϱ_b). Next

(i) A Cauchy sequence in X is also a Cauchy sequence in metric space (X, ϱ_b^s).
(ii) (X, ϱ_b) is a weak-partial b-metric space if (X, ϱ_b^s) is a complete metric space.
(iii) For any weak-partial b-metric ϱ_b on X, the function $\varrho_b^s : X \times X \to \mathbf{R}^+$ can be expressed as follows:

$$\varrho_b^s(\theta, \xi) = \varrho_b(\theta, \xi) - \frac{1}{2}[\varrho_b(\theta, \theta) + \varrho_b(\xi, \xi)],$$

which establishes a b-metric on X. Moreover, if $\{\theta_n\}$ in (X, ϱ_b^s) converges, it will eventually end at $\theta \in X$.

$$\lim_{n,m\to\infty} \varrho_b^s(\theta_n, \theta_m) = \lim_{n\to\infty} \varrho_b(\jmath_n, \jmath) = \varrho_b(\jmath, \jmath).$$

The multivalued partial metric space by Aydi et al. [3] was extended to weak-partial b-metric space by Kanwal et al. [17] in the following ways:

The class of all non-empty, closed, and bounded subsets of (X, ϱ_b) is $\mathcal{CB}^{\varrho_b}(X)$. Assume that the weak-partial b-metric space is (X, ϱ_b). Define: $\mathbf{W}, \mathbf{Z} \in \mathcal{CB}^{\varrho_b}(X)$, for $\theta \in X$.

$$\varrho_b(\theta, \mathbf{W}) = inf\{\varrho_b(\theta, \sigma) : \sigma \in \mathbf{W}\};$$
$$\delta(\mathbf{W}, \mathbf{Z}) = \sup\{\varrho_b(\sigma, \mathbf{Z}) : \sigma \in \mathbf{W}\};$$
$$\delta(\mathbf{Z}, \mathbf{W}) = \sup\{\varrho_b(\eta, \mathbf{W}) : \eta \in \mathbf{Z}\}.$$

Note that

$$\varrho_b(\theta, \mathbf{W}) = 0 \Longleftrightarrow \varrho_b^s(\theta, \mathbf{W}) = 0, \tag{1}$$

where

$$\varrho_b^s(\theta, \mathbf{W}) = \inf\{\varrho_b^s(\theta, \mathbf{W}), \theta \in \mathbf{W}\}. \tag{2}$$

Remark 1 ([17]) Assume that (X, ϱ_b) is a weak-partial b-metric space and $\mathbf{M}$ a non-empty subset of X, then

$$\theta \in \bar{\mathbf{W}} \Leftrightarrow \varrho_b(\theta, \mathbf{W}) = \varrho_b(\theta, \theta). \tag{3}$$

Definition 4 ([17]) Assume that the weak-partial b-metric space is (X, ϱ_b). The mapping $\mathcal{Z} : \mathcal{CB}^{\varrho_b} \times \mathcal{CB}^{\varrho_b} \to [0, \infty)$ for $\mathbf{W}, \mathbf{Z} \in \mathcal{CB}^{\varrho_b}(X)$ is defined by

$$\mathcal{H}(\mathbf{W}, \mathbf{Z}) = \frac{1}{2}\{\delta(\mathbf{W}, \mathbf{Z}) + \delta(\mathbf{W}, \mathbf{Z})\}, \tag{4}$$

is called $\mathcal{H}$-type Hausdorff metric induced by ϱ_b.

Proposition 1 ([17]) *Assume that (X, ϱ_b) is a weak-partial b-metric space. For $\mathbf{W}, \mathbf{Z}, \mathbf{Q} \in \mathcal{CB}^{\varrho_b}(X)$, we have*

(WHB1) $\mathcal{H}(\mathbf{W}, \mathbf{W}) \le \mathcal{Z}(\mathbf{W}, \mathbf{Z})$,
(WHB2) $\mathcal{H}(\mathbf{W}, \mathbf{Z}) = \mathcal{H}(\mathbf{Z}, \mathbf{W})$, *and*
(WHB3) $\mathcal{H}(\mathbf{W}, \mathbf{Z}) \le s[\mathcal{H}(\mathbf{W}, \mathbf{Q}) + \mathcal{H}(\mathbf{Q}, \mathbf{Z})]$.

Lemma 1 ([17]) *Assume that (X, ϱ_b) is a weak-partial b-metric space with $s \ge 1$. A multivalued mapping is represented by $f : X \to \mathcal{CB}^{\varrho_b}(X)$. Given a sequence in X, $\{\theta_n\}$, such that $\theta_n \in f\theta_{n-1}$, and*

$$\varrho_b(\theta_n, \theta_{n+1}) \le k\varrho_b(\theta_{n-1}, \theta_n), \tag{5}$$

for each $\theta \in X$, where $k \in (0, 1)$, then $\{\theta_n\}$ is Cauchy.

Definition 5 ([17]) Let (X, ϱ_b) be a complete weak-partial b-metric spaces. Let $f : X \to \mathcal{CB}^{\varrho_b}(X)$ be a multivalued $\mathcal{H}$-contraction mapping if

(1′) for every $\theta, \xi \in X, \exists\, k \in (0, 1)$ such that

$$+\,\mathcal{H}(f\theta \backslash \{\theta\}, f\xi \backslash \{\xi\}) \leq k\varrho_b(\theta, \xi), \tag{6}$$

(2′) for every $\theta \in X, \xi \in f\theta$ and $\epsilon > 0, \exists\, \jmath \in f\xi$ such that

$$\varrho_b(\theta, \jmath) \leq \mathcal{H}(f\theta, f\xi) + \epsilon. \tag{7}$$

Using the concepts of Jungck [15] and Sessa [32], we introduce the following definitions:

Definition 6 If $f\theta = g\theta = \theta^*$, then a point $\theta^* \in X$ on a pair of multivalued mappings (f, g) on a weak-partial b-metric space (X, ϱ_b) is termed the coincidence point of (f, g). θ is considered to be a common fixed point if $\theta^* = \theta$.

Definition 7 Let (f, g) be a pair of two multivalued mappings on X, and let (X, ϱ_b) be a weak-partial b-metric space. Then

(i) $g(f\theta) = f(g\theta)$ is a commuting mapping on X for any $\theta \in X$.
(ii) For all $\varrho_b(g(f\theta), f(g\theta)) \leq \varrho_b(g\theta, f\theta)$ are said to be weakly commuting.
(iii) When θ_{2n} is a sequence in X, we obtain for compatible mapping:
  (a) $\lim_{n\to\infty} \varrho_b(gf\theta_{2n}, fg\theta_{2n}) = 0$,
  (b) $\lim_{n\to\infty} f\theta_{2n} = \lim_{n\to\infty} g\theta_{2n} = \theta^*$.
(iv) For all $g(f\theta) = gf(g\theta)$ are said to be weakly compatible, for every coincidence point $\theta \in X$.

We extend the following lemma in [13] from b-metric space notion to weak-partial b-metric space concept.

Lemma 2 *Assume that $s \geq 1$ and that (X, ϱ_b, s) is a weak-partial b-metric space. Assume that $\{\theta_n\}$ and $\{\xi_n\}$ are sequences that converge to ϱ_b for θ and ξ, respectively. Next,*

$$\frac{1}{s}\varrho_b(\theta, \xi) \leq \lim_{n\to+\infty} \inf \varrho_b(\theta_n, \xi_n) \leq \lim_{n\to+\infty} \sup \varrho_b(\theta_n, \xi_n) \leq s^2\varrho_b(\theta, \xi).$$

In case $\theta = \xi$, we get

$$\lim_{n\to+\infty} \varrho_b(\theta_n, \xi_n) = 0.$$

Moreover for each $\theta \in X$,

$$\frac{1}{s}\varrho_b(\theta,\xi) \leq \lim_{n\to+\infty} \inf \varrho_b(\theta_n,\xi) \leq \lim_{n\to+\infty} \sup \varrho_b(\theta_n,\xi) \leq s^2\varrho_b(\theta,\xi).$$

Remark 2 ([12]) A topological space is T_1 if, given two distinct point's $\theta_1, \theta_2 \in X$, both points have a neighborhood that does not contain the other.

In 1944, Dieudonné [12] introduced the concept of paracompact space by showing that it is weaker than the concept of compact space.

Definition 8 ([12, 35]) X is said to be paracompact if it is Hausdorff and for each open covering $\mathbf{R}$ of X, there exists a locally finite open $\mathbf{R}$ that is finer than $\mathbf{R}$.

Theorem 1 ([23]) *If the mapping $f : X \to \mathbf{F}(\mathcal{N})$ is convex-valued lower semi-continuous (l.s.c.), with X paracompact space and $\mathcal{N}$ a Banach space, then f allows a (single-valued) continuous selection.*

Let $\mathcal{H}$ be the Hausdorff metric on $P_{cl,bd}(X)$ induced by d; that is,

$$\mathcal{H}(\mathbf{W},\mathbf{Z}) = \max\{\sup_{\theta\in\mathbf{Z}} \delta(\theta,\mathbf{W}), \sup_{\theta\in\mathbf{W}} \delta(\theta,\mathbf{Z})\}, for\ \mathbf{W},\mathbf{Z} \in P_{cl,bd}(X).$$

Then $\delta(\theta,\mathbf{W}) = \{\sup \delta(\theta,\xi) : \xi \in \mathbf{W}\}$ and $\delta(\theta,\mathbf{W}) = \inf d(\theta,\xi) : \xi \in \mathbf{W}$ where $X \in P_{cl,bd}(X)$. We denote $p_{cl}(X)$ the class of all non-empty, closed, and bounded subsets of X. CL-closed sets in X.

Khojasteh et al. [18] proved the theorem given below for multivalued mapping in complete metric space.

Theorem 2 ([18]) *The following requirements apply if (X, d) is a complete metric space and f is a multivalued mapping from X into $\mathcal{CB}(X)$,*

$$\mathcal{H}(f\theta, f\xi) \leq \frac{D(\xi,f\theta) + D(\theta,f\xi)}{\delta(\theta,f\theta) + \delta(\xi,f\xi)} d(\theta,\xi), \tag{8}$$

for all $\theta, \xi \in X$. Then, f has a fixed point $\jmath \in X$.

A generalized version of the preceding theorem for two multivalued mappings in complete metric spaces was provided by Rhoades [29].

Theorem 3 ([29]) *Let X be a complete metric spaces, $f, g : X \to \mathcal{CL}(X)$ satisfying, for all $\theta, \xi \in X$,*

$$\mathcal{H}(g\theta, f\xi) \leq \mathcal{N}(\theta,\xi) m(\theta,\xi), \tag{9}$$

where

$$\mathcal{N}(\theta,\xi) := \frac{\max\{d(\theta,\xi), D(\theta,f\theta) + D(\xi,g\xi), D(\theta,g\xi) + D(\xi,f\theta)\}}{\delta(\theta,f\theta) + \delta(\xi,g\xi) + 1} \tag{10}$$

and

$$m(\theta, \xi) := \max\left\{d(\theta, \xi), D(\theta, f\theta) + D(\xi, g\xi), \frac{D(\theta, g\xi) + D(\xi, f\theta)}{2}\right\}. \quad (11)$$

Then

(*a*) *There is at least one common fixed point* $p \in \mathcal{X}$ *between* f *and* g.
(*b*) *Each* $\theta \in \mathcal{X}$ *has a common fixed point that is reached by* $\{(fg)^{\frac{n}{2}}\theta\}$ *and* $\{g(fg)^{\frac{n}{2}}\theta\}$ *for n even.*
(c) $d(p, q) \geq \frac{1}{2}$ *if p and q are separate common fixed points of f and g.*

Lemma 3 ([29]) *Given a complete metric space* (X, d) *and a sequence* $\{\theta_n\}$ *in* X*, allow*

$$d(\theta_n, \theta_{n+1}) \leq \beta_n d(\theta_{n-1}, \theta_n), n = 1, 2, \ldots,$$

where

$$\beta_n = \frac{d(\theta_{n-1}, \theta_n) + d(\theta_n, \theta_{n+1})}{d(\theta_{n-1}, \theta_n) + d(\theta_n, \theta_{n+1}) + 1}.$$

Then $\{\theta_n\}$ *is a Cauchy sequence.*

Furthermore, by providing a proof of the following theorem, Demma and Vetro [11] extended Theorem 2 from metric space to *b*-metric space.

Theorem 4 ([11]) *Let* $f : X \to \mathcal{CB}(X)$ *be a mapping such that, given a complete b-metric space* (X, d, s)*, we have*

$$s\mathcal{H}(f\theta, f\xi) \leq \frac{d(\theta, f\xi) + d(\xi, f\theta)}{d(\theta, f\theta) + d(\xi, f\xi) + 3a} \quad (12)$$

for all $\theta, \xi \in X$ *and* $s \geq 1$*, where a is a positive real number. Then, f has a fixed point* $\jmath \in X$.

The following common fixed point theorem for two multivalued mappings in complete *b*-metric spaces was recently proved by Aliouche and Hamaizia [2].

Theorem 5 *Let* (X, d) *be a full b-metric space where* $f, g : X \to P_{cl}(X)$ *fulfilling and d continuous.*

$$s^{\alpha}\mathcal{H}(f\theta, g\xi) \leq \mathcal{N}(\theta, \xi)M(\theta, \xi), \quad (13)$$

where

$$\mathcal{N}(\theta, \xi) : \leq \frac{\max d(\theta, \xi), D(\theta, f\theta) + D(\xi, g\xi), D(\theta, g\xi) + D(\xi, fv) + \tau}{\delta(\theta, f\theta) + \delta(\xi, g\xi) + \gamma} \quad (14)$$

$\gamma > \tau \geq 0$ *and*

$$M(\theta,\xi):\ \leq \max\left\{d(\theta,\xi), D(\theta,f\theta)+D(\xi,g\xi), \frac{D(\theta,g\xi)+D(\xi,f\theta)}{2s}\right\} \quad (15)$$

for all $\theta,\xi \in X$, $s \geq 1$. *Then*

(*a*) There is at least one fixed point $\jmath \in X$ shared by f and g.
(*b*) Each $\theta \in X$ has a common fixed point that $\{(fg)^{\frac{n}{2}}\theta\}$ and $\{g(fg)^{\frac{n}{2}}\theta\}$ converge to for every n even.
(*c*) $d(\jmath, w) \geq \frac{\gamma s^{\alpha}-\beta}{2}$ if $f\jmath = g\jmath = \{\jmath\}$ and w is another separate common fixed point of f and g.

3 Main Results

Using two multivalued mappings in weak-partial b-metric space, we present the following lemma.

Lemma 4 *Let* (X, ϱ_b) *be weak-partial b-metric space with* $s \geq 1$ *and* $f, g : X \to P_{cl}^{\varrho_b}(X)$ *be a multivalued mapping with* X *paracompact space. If* $\{\theta_n\}$ *is a sequence in* X *such that* $\theta_{2n} \in f\theta_{2n-1}$ *and* $\theta_{2n+1} \in g\theta_{2n}$

$$\varrho_b(\theta_{2n},\theta_{zn+1}) \leq \varrho_b(f\theta_{2n-1}, g\theta_{2n}) \leq k\varrho_b(\theta_{2n-1},\theta_{2n}), \quad (16)$$

for each $\theta \in X$, *where* $k \in (0,1)$, *then* $\{\theta_n\}$ *is a Cauchy sequence.*

We use two multivalued mapping in weak-partial b-metric space setup to enhance Definition 5 as follows:

Definition 9 Let (X, ϱ_b) represent a weak-partial complete b-metric spaces. A multivalued mapping $f, g : X \to P_{cl}^{\varrho_b}(X)$ is known as $\mathcal{H}$-contraction with X paracompact space if

(1′) There exists $k \in (0,1)$ such that, for each $\theta,\xi \in X$,

$$s^{\alpha}\mathcal{H}(f\theta\backslash\{\theta\}, g\xi\backslash\{\xi\}) \leq k\mathcal{Q}(\theta,\xi), \quad (17)$$

where

$$k = \mathcal{N}(\theta,\xi) := \frac{\max\left\{\varrho_b(\theta,\xi), \varrho_b(\theta,f\theta)+\varrho_b(\xi,g\xi), \varrho_b(\theta,g\xi)+\varrho_b(\xi,f\xi)+\tau\right\}}{\varrho_b(\theta,f\theta)+\varrho_b(\xi,g\xi)+\gamma} \quad (18)$$

and

$$\mathcal{Q}(\theta,\xi) := \max\left\{\varrho_b(\theta,\xi), \varrho_b(\theta,g\theta), \varrho_b(\xi,f\xi)\right\} \quad (19)$$

for all $\theta, \xi \in X$, $s \geq 1$. As a result, every fixed point of f is a reversal and a fixed point of g.

(2′) For every $\theta \in X$, $\xi \in f\theta$ and $\epsilon > 0$, $\exists\, \jmath \in g\xi$ such that

$$\varrho_b(\xi, \jmath) \leq \mathcal{H}(f\theta, g\xi) + \epsilon. \tag{20}$$

$P_{cl}^{\varrho_b}$ represents the class of all closed non-empty subsets of X.

We establish the subsequent theorem.

Theorem 6 *Let (X, ϱ_b) be a complete weak-partial b-metric space, and let $f, g : X \to P_{cl}^{\varrho_b}(X)$ be a pair of multivalued mappings in a paracompact space, that fulfill inequalities (17), (18), (19), and (20). Then*

(a) *There is at least one fixed point shared by f and g. $\jmath \in X$,*

(b) *Each $\theta \in X$ has a common fixed point that $\{(fg)^{\frac{n}{2}}\theta\}$ and $\{g(fg)^{\frac{n}{2}}\theta\}$ converge to for every n even.*

(c) *$\varrho_b(\jmath, \jmath^*) \geq \frac{\gamma s^{2\alpha} - \tau}{2}$ is the result iff $\jmath = g\jmath = \{\jmath\}$ and $\jmath, \jmath^*$ are distinct strict common fixed points of f and g satisfying a T_1 space in X. Since X is a paracompact space, f and g satisfy a T_1 space in X.*

Proof Suppose θ_0 is an arbitrary point in X. According to Definition 8, X is a paracompact space. We define a sequence $\{\theta_{2n}\}$ in X such that for all $n \in \mathbf{N}$ and $0 < k < 1$, we have $\theta_{2n+1} \in f\theta_{2n} = g\theta_{2n+1}$ and $\theta_{2n+2} \in f\theta_{2n+1} = g\theta_{2n+2}$. This is equivalent to $\theta_{2n+1} \in (fg^{\frac{n}{2}})\theta$ and $\theta_{2n+2} \in (fg^{\frac{n}{2}})\theta$, which implies that $\mathcal{H}(f\theta_{2n}, g\theta_{2n+1}) = \mathcal{H}(f\theta_{2n+1}, g\theta_{2n+2})$. When $\mathcal{H}(f\theta_{2n}, g\theta_{2n+1}) = 0$ is reached for a given value of n, $\theta_{2n+1} \in g\theta_{2n+1}$, and therefore $\theta_{2n+1} \in F(g)$. $\theta_{2n} \in F(f)$, $\theta_{2n} = \theta_{2n+1}$, and θ_{2n} is stated to be a common fixed point using (20). So, (a) is confirmed.

Likewise, if $\mathcal{H}(f\theta_{2n+1}, g\theta_{2n+2}) = 0$ is satisfied for some value of n, then $\theta_{2n+1} \in F(f) \cap F(g)$ and (a) is confirmed once more.

Therefore, we assume that $\mathcal{H}(f\theta_{2n}, g\theta_{2n+1}) \neq 0$ and $\mathcal{H}(f\theta_{2n+1}, g\theta_{2n+2}) \neq 0$ for all $n \in \mathbf{N}$.

Let $\theta = \theta_{2n} \in f\theta_{2n}$. From (20), choose $\xi = \theta_{2n+1} \in g\theta_{2n+1}$ such that

$$\varrho_b(\theta_{2n}, \theta_{2n+1}) \leq \mathcal{H}(f\theta_{2n-1}, g\theta_{2n}) + \epsilon. \tag{21}$$

Assume that $\epsilon = \left(\frac{1}{\sqrt{k}} - 1\right)\mathcal{H}(f\theta_{2n-1}, g\theta_{2n})$ in (21), we have

$$\begin{aligned}
\varrho_b(\theta_1, \theta_2) &\leq \mathcal{H}(f\theta_0, g\theta_1) + \left(\frac{1}{\sqrt{k}} - 1\right)\mathcal{H}^+_{\varrho_b}(f\theta_0, g\theta_1), \\
&\leq \mathcal{H}(f\theta_{2n}, g\theta_{2n+1}) + \frac{1}{\sqrt{k}}\mathcal{H}(f\theta_{2n-1}, g\theta_{2n}) - \mathcal{H}(f\theta_{2n-1}, g\theta_{2n}), \\
&\leq \frac{1}{\sqrt{k}}\mathcal{H}(f\theta_{2n-1}, g\theta_{2n}), \ \textit{for}\ 0 < k < 1.
\end{aligned} \tag{22}$$

In similar manner, there exists $\theta_{2n+1} \in f\theta_{2n+1}$ and $\theta_{2n+2} \in g\theta_{2n+2}$ such that

$$\varrho_b(\theta_{2n+1}, \theta_{2n+2}) \leq \frac{1}{\sqrt{k}} \mathcal{H}(f\theta_{2n}, g\theta_{2n+1}), \forall\, 0 < k < 1. \tag{23}$$

Proceeding this same way through induction, by using (17) in (23), we have

$$\begin{aligned} \varrho_b(\theta_{2n+1}, \theta_{2n+2}) &\leq \frac{1}{s^\alpha} \frac{1}{\sqrt{k}} k \mathcal{Q}(\theta_{2n}, \theta_{2n+1}), \\ &\leq \frac{\sqrt{k}}{s^\alpha} \mathcal{Q}(\theta_{2n}, \theta_{2n+1}), \end{aligned} \tag{24}$$

where $k = \mathcal{N}(\theta_{2n}, \theta_{2n+1}) = s\beta_{2n}$, and $s\beta_{2n}$ is generated by using (18) as follows:

$$\begin{aligned} \mathcal{N}(\theta_{2n}, \theta_{2n+1}) &\leq \frac{\max \left\{ \begin{array}{r} \varrho_b(\theta_{2n}, \theta_{2n+1}), \varrho_b(\theta_{2n}, f\theta_{2n}) + \varrho_b(\theta_{2n+1}, g\theta_{2n+1}), \\ \varrho_b(\theta_{2n}, g\theta_{2n+1}) + \varrho_b(\theta_{2n+1}, f\theta_{2n}) + \tau \end{array} \right\}}{\varrho_b(\theta_{2n}, f\theta_{2n}) + \varrho_b(\theta_{2n+1}, g\theta_{2n+1}) + \gamma}, \\ &\leq \frac{\max \left\{ \begin{array}{r} \varrho_b(\theta_{2n}, \theta_{2n+1}), \varrho_b(\theta_{2n}, \theta_{2n+1}) + \varrho_b(\theta_{2n+1}, \theta_{2n+2}), \\ \varrho_b(\theta_{2n}, \theta_{2n+2}) + \varrho_b(\theta_{2n+1}, \theta_{2n+1}) + \tau \end{array} \right\}}{\varrho_b(\theta_{2n}, \theta_{2n+1}) + \varrho_b(\theta_{2n+1}, \theta_{2n+2}) + \gamma}. \end{aligned}$$

By $(WPB4)$ in Definition 1 and we get

$$\begin{aligned} \mathcal{N}(\theta_{2n}, \theta_{2n+1}) &\leq \frac{\max \left\{ \begin{array}{r} \varrho_b(\theta_{2n}, \theta_{2n+1}), \varrho_b(\theta_{2n}, \theta_{2n+1}) + \varrho_b(\theta_{2n+1}, \theta_{2n+2}), \\ s[\varrho_b(\theta_{2n}, \theta_{2n+1}) + \varrho_b(\theta_{2n+1}, \theta_{2n+2})] + \\ \varrho_b(\theta_{2n+1}, \theta_{2n+1}) + \tau \end{array} \right\}}{\varrho_b(\theta_{2n}, \theta_{2n+1}) + \varrho_b(\theta_{2n+1}, \theta_{2n+2}) + \gamma}, \\ &\leq \frac{\max \left\{ \begin{array}{r} \varrho_b(\theta_{2n}, \theta_{2n+1}), \varrho_b(\theta_{2n}, \theta_{2n+1}) + \varrho_b(\theta_{2n+1}, \theta_{2n+2}), \\ s[\varrho_b(\theta_{2n}, \theta_{2n+1}) + \varrho_b(\theta_{2n+1}, \theta_{2n+2})] + \tau \end{array} \right\}}{\varrho_b(\theta_{2n}, \theta_{2n+1}) + \varrho_b(\theta_{2n+1}, \theta_{2n+2}) + \gamma}, \end{aligned}$$

which gives

$$\mathcal{N}(\theta_{2n}, \theta_{2n+1}) \leq \frac{s[\varrho_b(\theta_{2n}, \theta_{2n+1}) + \varrho_b(\theta_{2n+1}, \theta_{2n+2})] + \tau}{\varrho_b(\theta_{2n}, \theta_{2n+1}) + \varrho_b(\theta_{2n+1}, \theta_{2n+2}) + \gamma}. \tag{25}$$

Applying the concept of Lemma 3, $\gamma > \tau \geq 0$ and define $\varrho_{b_{2n}} = \varrho_b(\theta_{2n}, \theta_{2n+1})$ we obtain

$$\begin{aligned} \mathcal{N}(\theta_{2n}, \theta_{2n+1}) &\leq \frac{s[\varrho_{b_{2n}} + \varrho_{b_{2n+1}}] + \tau}{\varrho_{b_{2n}} + \varrho_{b_{2n+1}} + \gamma}, \\ &= s\beta_{2n+1}. \end{aligned} \tag{26}$$

From (19)

$$
\begin{aligned}
\mathcal{Q}(\theta_{2n}, \theta_{2n+1}) &\leq \max\Big\{\varrho_b(\theta_{2n}, \theta_{2n+1}), \varrho_b(\theta_{2n}, g\theta_{2n}), \varrho_b(\theta_{2n+1}, f\theta_{2n+1})\Big\},\\
&\leq \max\Big\{\varrho_b(\theta_{2n}, \theta_{2n+1}), \varrho_b(\theta_{2n}, \theta_{2n+1}), \varrho_b(\theta_{2n+1}, \theta_{2n+2})\Big\},\\
&\leq \max\Big\{\varrho_b(\theta_{2n}, \theta_{2n+1}), \varrho_b(\theta_{2n+1}, \theta_{2n+2})\Big\}, \forall\, n = 0, 1, \ldots
\end{aligned}
$$

If $\mathcal{Q}(\theta_{2n}, \theta_{2n+1}) = \varrho_b(\theta_{2n+1}, \theta_{2n+2})$, then

$$
\begin{aligned}
\varrho_b(\theta_{2n+1}, \theta_{2n+2}) &\leq \frac{\sqrt{k}}{s^{\alpha}}\mathcal{Q}(\theta_{2n}, \theta_{2n+1}),\\
&\leq \frac{\sqrt{k}}{s^{\alpha}}\varrho_b(\theta_{2n+1}, \theta_{2n+2}),
\end{aligned}
$$

which is a contradiction. Hence, we have $\mathcal{Q}(\theta_{2n}, \theta_{2n+1}) = \varrho_b(\theta_{2n}, \theta_{2n+1})$. By substituting (26) in (24) we have

$$
\begin{aligned}
\varrho_b(\theta_{2n+1}, \theta_{2n+2}) &\leq \frac{\sqrt{k}}{s^{\alpha}}\mathcal{Q}(\theta_{2n}, \theta_{2n+1}),\\
&\leq \frac{\sqrt{k}}{s^{\alpha}}\varrho_b(\theta_{2n}, \theta_{2n+1}),\\
&\leq \frac{\sqrt{s\beta_{2n+1}}}{s^{\alpha}}\varrho_b(\theta_{2n}, \theta_{2n+1}).
\end{aligned}
$$

Equivalently to Lemma 4

$$
\varrho_b(\theta_{2n+1}, \theta_{2n+2}) \leq k\varrho_b(\theta_{2n}, \theta_{2n+1}),
$$

where $\theta = \frac{\sqrt{s\beta_{2n+1}}}{s^{\alpha}}$.

By Lemma 3, $\{\theta_{2n}\}$ is a Cauchy sequence and so is $\{\theta_n\}$.

Using (17) in (23) for $\mathcal{H}^{+}_{\varrho_b}(f\theta_{2n}, g\theta_{2n-1}) = 0$, we have

$$
\begin{aligned}
\varrho_b(\theta_{2n}, \theta_{2n+1}) &\leq \frac{1}{\sqrt{k}}\mathcal{H}(f\theta_{2n}, g\theta_{2n-1}),\\
&\leq \frac{1}{s^{\alpha}}\frac{1}{\sqrt{k}}k\mathcal{Q}(\theta_{2n}, \theta_{2n-1}),\\
&\leq \frac{\sqrt{k}}{s^{\alpha}}\mathcal{Q}(\theta_{2n}, \theta_{2n-1}),
\end{aligned}
\tag{27}
$$

where $k = \mathcal{N}(\theta_{2n}, \theta_{2n-1}) = s\beta_{2n}$, and $s\beta_{2n}$ is generated by using (76) as follows:

$$\begin{aligned}
\mathcal{N}(\theta_{2n}, \theta_{2n-1}) &\leq \frac{\max\left\{\begin{array}{r} \varrho_b(\theta_{2n}, \theta_{2n-1}), \varrho_b(\theta_{2n}, f\theta_{2n-1}) + \varrho_b(\theta_{2n}, g\theta_{2n-1}), \\ \varrho_b(\theta_{2n}, g\theta_{2n-1}) + \varrho_b(\theta_{2n-1}, f\theta_{2n}) + \tau \end{array}\right\}}{\varrho_b(\theta_{2n}, f\theta_{2n}) + \varrho_b(\theta_{2n-1}, g\theta_{2n-1}) + \gamma}, \\
&\leq \frac{\max\left\{\begin{array}{r} \varrho_b(\theta_{2n}, \theta_{2n-1}), \varrho_b(\theta_{2n}, \theta_{2n}) + \varrho_b(\theta_{2n}, \theta_{2n}), \\ \varrho_b(\theta_{2n}, \theta_{2n}) + \varrho_b(\theta_{2n-1}, \theta_{2n}) + \tau \end{array}\right\}}{\varrho_b(\theta_{2n}, \theta_{2n+1}) + \varrho_b(\theta_{2n-1}, \theta_{2n}) + \gamma}. \qquad (28)
\end{aligned}$$

Using $(WPB4)$ in Definition 1 we get

$$\begin{aligned}
\mathcal{N}(\theta_{2n}, \theta_{2n-1}) &\leq \frac{\max\left\{\varrho_b(\theta_{2n}, \theta_{2n-1}), \varrho_b(\theta_{2n-1}, \theta_{2n+1}) + \tau\right\}}{\varrho_b(\theta_{2n}, \theta_{2n+1}) + \varrho_b(\theta_{2n-1}, \theta_{2n}) + \gamma}, \\
&\leq \frac{\max\left\{\varrho_b(\theta_{2n}, \theta_{2n-1}), s[\varrho_b(\theta_{2n-1}, \theta_{2n}) + \varrho_b(\theta_{2n}, \theta_{2n+1})] + \tau\right\}}{\varrho_b(\theta_{2n}, \theta_{2n+1}) + \varrho_b(\theta_{2n-1}, \theta_{2n}) + \gamma},
\end{aligned}$$

which gives

$$\mathcal{N}(\theta_{2n}, \theta_{2n-1}) \leq \frac{s[\varrho_b(\theta_{2n-1}, \theta_{2n}) + \varrho_b(\theta_{2n}, \theta_{2n+1})] + \tau}{\varrho_b(\theta_{2n}, \theta_{2n+1}) + \varrho_b(\theta_{2n-1}, \theta_{2n}) + \gamma}. \qquad (29)$$

Applying the concept of Lemma 3 and define $\varrho_{b_{2n}} = \varrho_b(\theta_{2n}, \theta_{2n+1})$ we obtain

$$\begin{aligned}
\mathcal{N}(\theta_{2n}, \theta_{2n-1}) &\leq \frac{s[\varrho_{b_{2n-1}} + \varrho_{b_{2n}}] + \tau}{\varrho_{b_{2n}} + \varrho_{b_{2n-1}} + \gamma}, \\
&= s\beta_{2n}. \qquad (30)
\end{aligned}$$

By Eq. (19), we have

$$\begin{aligned}
\mathcal{Q}(\theta_{2n}, \theta_{2n-1}) &\leq \max\left\{\varrho_b(\theta_{2n}, \theta_{2n-1}), \varrho_b(\theta_{2n}, g\theta_{2n}), \varrho_b(\theta_{2n-1}, f\theta_{2n-1})\right\}, \\
&\leq \max\left\{\varrho_b(\theta_{2n}, \theta_{2n-1}), \varrho_b(\theta_{2n}, \theta_{2n+1}), \varrho_b(\theta_{2n-1}, \theta_{2n})\right\}, \\
&\leq \max\left\{\varrho_b(\theta_{2n}, \theta_{2n-1}), \varrho_b(\theta_{2n}, \theta_{2n+1})\right\}, \forall\, n = 0, 1, \ldots
\end{aligned}$$

If $\mathcal{Q}(\theta_{2n}, \theta_{2n-1}) = \varrho_b(\theta_{2n}, \theta_{2n+1})$, then

$$\begin{aligned}
\varrho_b(\theta_{2n}, \theta_{2n+1}) &\leq \frac{\sqrt{k}}{s^{\alpha}} \mathcal{Q}(\theta_{2n}, \theta_{2n+1}), \\
&\leq \frac{\sqrt{k}}{s^{\alpha}} \varrho_b(\theta_{2n}, \theta_{2n+1}),
\end{aligned}$$

which is a contradiction. Hence, we have $\mathcal{Q}(\theta_{2n}, \theta_{2n-1}) = \varrho_b(\theta_{2n}, \theta_{2n-1})$. By substituting (30) in (24) we have

$$\begin{aligned}\varrho_b(\theta_{2n}, \theta_{2n+1}) &\leq \frac{\sqrt{k}}{s^\alpha}\mathcal{Q}(\theta_{2n}, \theta_{2n-1}),\\ &\leq \frac{\sqrt{k}}{s^\alpha}\varrho_b(\theta_{2n}, \theta_{2n-1}),\\ &\leq \frac{\sqrt{s\beta_{2n}}}{s^\alpha}\varrho_b(\theta_{2n}, \theta_{2n-1}).\end{aligned}$$

Equivalently to Lemma 4

$$\varrho_b(\theta_{2n}, \theta_{2n+1}) \leq k\varrho_b(\theta_{2n}, \theta_{2n-1}),$$

where $k = \frac{\sqrt{s\beta_{2n}}}{s^\alpha}$.

Both $\{\theta_{2n}\}$ and $\{\theta_n\}$ are Cauchy sequences according to Lemma 3. By applying (iii) from Definition 3 and the definition of ϱ_b^s, we derive the following: as m, n tends to $+\infty$, $\varrho_b(\theta_n, \theta_m)$ tends to zero. This suggests that $\{\theta_n\}$ is Cauchy in b-metric space (X, ϱ_b^s). (X, ϱ_b) is a complete b-metric space since (X, ϱ_b) is complete. Consequently, we have

$$\lim_{n,m\to\infty} \varrho_b^s(\theta_n, \theta_m) = \lim_{n\to\infty} \varrho_b(\jmath_n, \jmath) = \varrho_b(\jmath, \jmath).$$

Assuming that $\{\theta_{2n}\}$ is not a Cauchy sequence in X, there is $\epsilon > 0$ and two sequences of positive integers, $\{m_p\}$ and $\{n_p\}$, such that the following two sequences tend to ϵ when $p \to \infty$: $\varrho_b(\theta_{2m_p}, \theta_{2n_p})$ and $\varrho_b(\theta_{2m_p}, \theta_{2n_{p+1}})$. It can be demonstrated that the two provided sequences are Cauchy sequences in ϱ_b-metric space using Lemma 1. Suppose that

$$n_p > m_p, \varrho_b(\theta_{2m_p}, \theta_{2n_{p-1}}) < \epsilon, \varrho_b(p_{2m_p}, \theta_{2n_p}) > \epsilon, \lim_{p\to\infty} \varrho_b(\theta_{2m_p}, \kappa_{2m_p}) = 0. \quad (31)$$

From $(WPB4)$, we have

$$\begin{aligned}\epsilon \leq \varrho_b(\theta_{2m_p}, \theta_{2n_p}) &\leq s[\varrho_b(\theta_{2m_p}, \theta_{2n_{p-1}}) + \varrho_b(\theta_{2n_{p-1}}, \theta_{2n_p})],\\ &\leq s\varrho_b(\theta_{2m_p}, \theta_{2n_{p-1}}) + s\varrho_b(\theta_{2n_{p-1}}, \theta_{2n_p}),\\ &\leq s\epsilon + s\varrho_b(\theta_{2n_{p-1}}, \theta_{2n_p}),\\ \epsilon \leq \varrho_b(\theta_{2m_p}, \theta_{2n_p}) &\leq s\epsilon. \end{aligned} \quad (32)$$

Taking the lower limit as $p \to \infty$ in (32), and using (31) we obtain

$$\epsilon \leq \varrho_b(\theta_{2m_p}, \theta_{2n_p}) \leq \lim_{n\to+\infty} \inf \varrho_b(\theta_{2m_p}, \theta_{2n_p}) \leq s\epsilon. \quad (33)$$

Using $(WPB4)$, Lemma (1) and (33), we have

$$\begin{aligned}\epsilon \leq \varrho_b(\theta_{2m_p}, \theta_{2n_p}) &\leq s[\varrho_b(\theta_{2m_p}, \theta_{2n_{p+1}}) + \varrho_b(\theta_{2n_{p+1}}, \theta_{2n_p})],\\ &\leq s\varrho_b(\theta_{2m_p}, \theta_{2n_{p+1}}) + s\varrho_b(\theta_{2n_{p+1}}, \theta_{2n_p}),\\ &\leq s\varrho_b(\theta_{2m_p}, \theta_{2n_{p+1}}),\\ &\leq s^3\epsilon, \end{aligned} \tag{34}$$

and

$$\begin{aligned}\epsilon \leq \varrho_b(\theta_{2m_p}, \theta_{2n_{p+1}}) &\leq s[\varrho_b(\theta_{2m_p}, \theta_{2n_p}) + \varrho_b(\theta_{2n_p}, \theta_{2n_{p+1}})],\\ &\leq s\varrho_b(\theta_{2m_p}, \theta_{2n_p}) + s\varrho_b(\theta_{2n_p}, \theta_{2n_{p+1}}),\\ &\leq s\varrho_b(\theta_{2m_p}, \theta_{2n_p}),\\ &\leq s^2\epsilon. \end{aligned} \tag{35}$$

It follows from (34), (35), and (33)

$$\frac{\epsilon}{s} \leq \varrho_b(\theta_{2m_p}, \theta_{2n_{p+1}}) \leq \lim_{n\to+\infty} \inf \varrho_b(\theta_{2m_p}, \theta_{2n_{p+1}}) \leq s^2\epsilon.$$

Thus $\{\theta_n\}$ is a Cauchy sequence in (X, ϱ_b).

Next, we demonstrate that any fixed point of f is a fixed point of g if and only if f and g satisfy the requirements of Theorem 6. Assume that a fixed point of f is $\jmath$. Using $(WPB4)$, (20), and (17), we have

$$\begin{aligned}\varrho_b(\jmath, g\jmath) &\leq s[\varrho_b(\jmath, \theta_{2n+1}) + \varrho_b(\theta_{2n+1}, g\jmath)],\\ &\leq s\varrho_b(\jmath, \theta_{2n+1}) + s\mathcal{H}^+_{\varrho_b}(f\theta_{2n}, g\jmath). \end{aligned} \tag{36}$$

Let $\theta = \theta_{2n}$ and $\xi = \jmath$ in (18) we obtain

$$\mathcal{N}(\theta_{2n}, \jmath) \leq \frac{\max\left\{\begin{matrix}\varrho_b(\theta_{2n}, \jmath), \varrho_b(\theta_{2n}, f\theta_{2n}) + \varrho_b(\jmath, g\jmath),\\ \varrho_b(\theta_{2n}, g\jmath) + \varrho_b(\jmath, f\theta_{2n}) + \tau\end{matrix}\right\}}{\varrho_b(\theta_{2n}, f\theta_{2n}) + \varrho_b(\jmath, g\jmath) + \gamma}. \tag{37}$$

Letting $n \to \infty$ in (37) one gets

$$\begin{aligned}\mathcal{N}(\theta_{2n}, \jmath) &\leq \frac{\max\left\{\varrho_b(\jmath, \jmath), \varrho_b(\jmath, \jmath) + \varrho_b(\jmath, g\jmath), \varrho_b(\jmath, g\jmath) + \varrho_b(\jmath, \jmath) + \tau\right\}}{\varrho_b(\jmath, \jmath) + \varrho_b(\jmath, g\jmath) + \gamma},\\ &\leq \frac{\max\left\{\varrho_b(\jmath, g\jmath), \varrho_b(\jmath, g\jmath) + \tau\right\}}{\varrho_b(\jmath, g\jmath) + \gamma},\\ &\leq \frac{\varrho_b(\jmath, g\jmath) + \tau}{\varrho_b(\jmath, g\jmath) + \gamma}. \end{aligned} \tag{38}$$

From (19)

$$\begin{aligned}\mathcal{Q}(\theta_{2n}, \jmath) &\le \max\left\{\varrho_b(\theta_{2n}, \jmath), \varrho_b(\jmath_{2n}, g\theta_{2n}), \varrho_b(\jmath, f\jmath)\right\},\\ &\le \max\left\{\varrho_b(\jmath, \jmath), \varrho_b(\jmath, g\jmath), \varrho_b(\jmath, \jmath)\right\},\\ &= \varrho_b(\jmath, g\jmath). \end{aligned} \tag{39}$$

Using (38), (39) in (36)

$$\begin{aligned}\varrho_b(\jmath, g\jmath) &\le s\varrho_b(\jmath, \theta_{2n+1}) + s\mathcal{H}^+_{\varrho_b}(f\theta_{2n}, g\jmath),\\ &\le s\varrho_b(\jmath, \theta_{2n+1}) + s\frac{\sqrt{k}}{s^\alpha}\mathcal{Q}(\theta_{2n}, \jmath),\\ &\le s\varrho_b(\jmath, \jmath) + s\frac{\sqrt{k}}{s^\alpha}\varrho_b(\jmath, g\jmath),\\ &\le s^{1-\alpha}\sqrt{k}\varrho_b(\jmath, g\jmath),\\ &\le s^{1-\alpha}\sqrt{\frac{\varrho_b(\jmath, g\jmath)+\tau}{\varrho_b(\jmath, g\jmath)+\gamma}}\varrho_b(\jmath, g\jmath).\end{aligned}$$

If $\varrho_b(\jmath, g\jmath) = 0$, then the inequality above is satisfied. Therefore, $\jmath$ is a fixed point of g as well.

In order to demonstrate (c), let f and g have two unique common fixed points, $\jmath$ and $\jmath^*$, that fulfill T_1 space in X, where X is a paracompact space. Next, we obtain by using an inequality (24).

$$\begin{aligned}s^\alpha\varrho_b(\jmath, \jmath^*) &\le \mathcal{H}(f\jmath, g\jmath^*),\\ &\le \frac{\sqrt{k}}{s^\alpha}\mathcal{Q}(\jmath, \jmath^*).\end{aligned} \tag{40}$$

Since $k = \mathcal{N}(\jmath, \jmath^*)$, we have

$$\begin{aligned}\mathcal{N}(\jmath, \jmath^*) &= \frac{\max\left\{\varrho_b(\jmath, \jmath^*), \varrho_b(\jmath, f\jmath) + \varrho_b(\jmath^*, g\jmath^*), \varrho_b(\jmath, g\jmath^*) + \varrho_b(\jmath^*, f\jmath) + \tau\right\}}{\varrho_b(\jmath, f\jmath) + \varrho_b(\jmath^*, g\jmath^*) + \gamma},\\ &\le \frac{\max\left\{\varrho_b(\jmath, \jmath^*), \varrho_b(\jmath, \jmath) + \varrho_b(\jmath^*, \jmath^*), \varrho_b(\jmath, \jmath^*) + \varrho_b(\jmath^*, \jmath) + \tau\right\}}{\varrho_b(\jmath, \jmath) + \varrho_b(\jmath^*, \jmath^*) + \gamma},\\ &\le \frac{\max\left\{\varrho_b(\jmath, \jmath^*), \varrho_b(\jmath, \jmath^*) + \varrho_b(\jmath^*, \jmath) + \tau\right\}}{\gamma},\\ &= \frac{2\varrho_b(\jmath, \jmath^*) + \tau}{\gamma},\end{aligned} \tag{41}$$

and

$$\begin{aligned}\mathcal{Q}(\jmath, \jmath^*) &\leq \max\left\{\varrho_b(\jmath, \jmath^*), \varrho_b(\jmath, g\jmath), \varrho_b(\jmath^*, f\jmath^*)\right\},\\ &\leq \max\left\{\varrho_b(\jmath, \jmath^*), \varrho_b(\jmath, \jmath), \varrho_b(\jmath^*, \jmath^*)\right\},\\ &= \varrho_b(\jmath, \jmath^*).\end{aligned} \tag{42}$$

Taking (41), (42) in (43), one gets

$$\begin{aligned}s^{\alpha}\varrho_b(\jmath, \jmath^*) &\leq \sqrt{k}\mathcal{Q}(\jmath, \jmath^*),\\ s^{\alpha}\varrho_b(\jmath, \jmath^*) &\leq \sqrt{\frac{2\varrho_b(\jmath, \jmath^*) + \tau}{\gamma}}\varrho_b(\jmath, \jmath^*),\\ s^{\alpha} &\leq \sqrt{\frac{2\varrho_b(\jmath, \jmath^*) + \tau}{\gamma}},\\ s^{2\alpha} &\leq \frac{2\varrho_b(\jmath, \jmath^*) + \tau}{\gamma},\\ \gamma s^{2\alpha} &\leq 2\varrho_b(\jmath, \jmath^*) + \tau,\\ \gamma s^{2\alpha} - \tau &\leq 2\varrho_b(\jmath, \jmath^*),\\ \frac{\gamma s^{2\alpha} - \tau}{2} &\leq \varrho_b(\jmath, \jmath^*),\end{aligned}$$

which shows that $\varrho_b(\jmath, \jmath^*) \geq \frac{\gamma s^{2\alpha} - \tau}{2}$. Hence (c) is verified.

The outcomes of Theorem 6 are demonstrated in the example that follows.

Example 3 Let $X = \{0, 1\}$ and $\varrho_b : X \times X \to \mathbf{R}_+$ be defined as follows: for all $\theta, \xi \in X$ with $s = 2$, $\varrho_b(\theta, \xi) = \frac{1}{2}|\theta - \xi|^2 + \frac{1}{2}\max\{\theta, \xi\}$. Assume that the complete weak-partial b-metric space is (X, ϱ_b). Let

$$f\theta = \begin{cases}[0, \frac{\theta}{2}], & 0 \leq \theta < \frac{2}{3},\\ [0, -\frac{\theta}{2}] + 1, & \frac{2}{3} \leq \theta \leq 1,\end{cases}$$

and

$$g\theta = \begin{cases}[0, \frac{\theta}{2}], & \theta \neq \frac{1}{2},\\ [0, \kappa + \frac{1}{2}], & \theta = \frac{1}{2}.\end{cases}$$

Define $f, g : X \longrightarrow 2^{[0,1]}$, $X = f \cup g = [0, \frac{2}{3}] \cup [\frac{2}{3}, 1]$, and $P_{cl}^{\varrho_b}(X) = 2^{[0,1]}$.

By Remark 1, we have

$$\theta \in \bar{F} \Leftrightarrow \varrho_b(\theta, W) = \varrho_b(\theta, \theta). \tag{43}$$

Since

$$(\theta, \xi) \in \left\{(0,0), \left(0, \frac{1}{2}\right), \left(0, \frac{2}{3}\right), (0,1), \left(\frac{2}{3}, 0\right), \left(\frac{2}{3}, \frac{2}{3}\right), \left(\frac{2}{3}, 1\right), (1,0), \left(1, \frac{2}{3}\right), (1,1)\right\}$$

$\forall\, \theta, \xi \in X$.

For

$$\begin{aligned}
\theta \in \{\bar{0}\} &\Leftrightarrow \varrho_b(\theta, \{0\}) = \varrho_b(\theta, \theta),\\
&\Leftrightarrow \varrho_b(\theta, 0) = \frac{1}{2}(\theta - 0)^2 + \frac{1}{2}\max\{\theta, \theta\} = \varrho_b(\theta, \theta),\\
&\Leftrightarrow \varrho_b(\theta, 0) = \frac{1}{2}\theta^2 + \frac{1}{2}\theta = \frac{1}{2}\theta, for \theta = 0,\\
&\Leftrightarrow \theta \in \{0\}.
\end{aligned}$$

For

$$\begin{aligned}
\theta \in \overline{\left\{0, \frac{1}{2}\right\}} &\Leftrightarrow \varrho_b\left(\theta, \left\{0, \frac{1}{2}\right\}\right) = \varrho_b(\theta, \theta),\\
&\Leftrightarrow \min\left\{\varrho_b\left(\theta, 0\right), \varrho_b\left(\theta, \frac{1}{2}\right)\right\} = \varrho_b(\theta, \theta),
\end{aligned}$$

$$\begin{aligned}
&\Leftrightarrow \min\left\{\frac{1}{2}\theta^2 + \frac{1}{2}\theta, \frac{1}{2}\left(\theta - \frac{1}{2}\right)^2 + \frac{1}{2}\max\left(\theta, \frac{1}{2}\right)\right\} = \frac{1}{2}\theta,\\
&\Leftrightarrow \frac{1}{2}\left(\theta - \frac{1}{2}\right)^2 + \frac{1}{2}\max\left(\theta, \frac{1}{2}\right) = \frac{1}{2}\theta, for \theta = \frac{1}{2},\\
&\Leftrightarrow \theta \in \left\{0, \frac{1}{2}\right\}.
\end{aligned}$$

For

$$\begin{aligned}
\theta \in \overline{\left\{0, \frac{2}{3}\right\}} &\Leftrightarrow \varrho_b\left(\theta, \left\{0, \frac{2}{3}\right\}\right) = \varrho_b(\theta, \theta),\\
&\Leftrightarrow \min\left\{\varrho_b\left(\theta, 0\right), \varrho_b\left(\theta, \frac{2}{3}\right)\right\} = \varrho_b(\theta, \theta),\\
&\Leftrightarrow \min\left\{\frac{1}{2}\theta^2 + \frac{1}{2}\theta, \frac{1}{2}\left(\theta - \frac{2}{3}\right)^2 + \frac{1}{2}\max\left(\theta, \frac{2}{3}\right)\right\} = \frac{1}{2}\theta,
\end{aligned}$$

$$\Leftrightarrow \frac{1}{2}\left(\theta - \frac{2}{3}\right)^2 + \frac{1}{2}\max\left(\theta, \frac{2}{3}\right) = \frac{1}{2}\theta, for\theta = \frac{2}{3},$$

$$\Leftrightarrow \theta \in \left\{0, \frac{2}{3}\right\}.$$

For

$$\theta \in \overline{\left\{\frac{2}{3}, \frac{2}{3}\right\}} \Leftrightarrow \varrho_b\left(\theta, \left\{\frac{2}{3}, \frac{2}{3}\right\}\right) = \varrho_b(\theta, \theta),$$

$$\Leftrightarrow \min\left\{\varrho_b\left(\theta, \frac{2}{3}\right), \varrho_b\left(\theta, \frac{2}{3}\right)\right\} = \varrho_b(\theta, \theta),$$

$$\Leftrightarrow \min\left\{\frac{1}{2}\left(\theta - \frac{2}{3}\right)^2 + \frac{1}{2}\max\left(\theta, \frac{2}{3}\right), \frac{1}{2}\left(\theta - \frac{2}{3}\right)^2 + \frac{1}{2}\max\left(\theta, \frac{2}{3}\right)\right\} = \frac{1}{2}\theta,$$

$$\Leftrightarrow \frac{1}{2}\left(\theta - \frac{2}{3}\right)^2 + \frac{1}{2}\max\left(\theta, \frac{2}{3}\right) = \frac{1}{2}\theta, for\ \theta = \frac{2}{3},$$

$$\Leftrightarrow \theta \in \left\{\frac{2}{3}, \frac{2}{3}\right\}.$$

For

$$\theta \in \overline{\left\{\frac{2}{3}, 1\right\}} \Leftrightarrow \varrho_b\left(\theta, \left\{\frac{2}{3}, 1\right\}\right) = \varrho_b(\theta, \theta),$$

$$\Leftrightarrow \min\left\{\varrho_b\left(\theta, \frac{2}{3}\right), \varrho_b\left(\theta, 1\right)\right\} = \varrho_b(\theta, \theta),$$

$$\Leftrightarrow \min\left\{\frac{1}{2}\left(\theta - \frac{2}{3}\right)^2 + \frac{1}{2}\max\left(\theta, \frac{2}{3}\right), \frac{1}{2}\left(\theta - 1\right)^2 + \frac{1}{2}\max\left(\theta, 1\right)\right\} = \frac{1}{2}\theta,$$

$$\Leftrightarrow \frac{1}{2}\left(\theta - 1\right)^2 + \frac{1}{2}\max\left(\theta, 1\right) = \frac{1}{2}\theta, for\ \theta = 1,$$

$$\Leftrightarrow \theta \in \left\{\frac{2}{3}, 1\right\}.$$

For

$$\theta \in \left\{\overline{1,1}\right\} \Leftrightarrow \varrho_b\Big(\theta, \left\{1,1\right\}\Big) = \varrho_b(\theta,\theta),$$
$$\Leftrightarrow \min\left\{\varrho_b\Big(\theta,1\Big), \varrho_b\Big(\theta,1\Big)\right\} = \varrho_b(\theta,\theta),$$
$$\Leftrightarrow \min\left\{\frac{1}{2}\Big(\theta-1\Big)^2 + \frac{1}{2}\max\Big(\theta,1\Big), \frac{1}{2}\Big(\theta-1\Big)^2 + \frac{1}{2}\max\Big(\theta,1\Big)\right\} = \frac{1}{2}\theta,$$
$$\Leftrightarrow \frac{1}{2}\Big(\theta-1\Big)^2 + \frac{1}{2}\max\Big(\theta,1\Big) = \frac{1}{2}\theta, for\ \theta = 1,$$
$$\Leftrightarrow \theta \in \left\{1,1\right\}.$$

$$\theta \in \left\{\overline{0,1}\right\} \Leftrightarrow \varrho_b\Big(\theta, \left\{0,1\right\}\Big) = \varrho_b(\theta,\theta),$$
$$\Leftrightarrow \min\left\{\varrho_b\Big(\theta,0\Big), \varrho_b\Big(\theta,1\Big)\right\} = \varrho_b(\theta,\theta),$$
$$\Leftrightarrow \min\left\{\frac{1}{2}\Big(\theta-0\Big)^2 + \frac{1}{2}\max\Big(\theta,0\Big), \frac{1}{2}\Big(\theta-1\Big)^2 + \frac{1}{2}\max\Big(\theta,1\Big)\right\} = \frac{1}{2}\theta,$$
$$\Leftrightarrow \frac{1}{2}\Big(\theta-1\Big)^2 + \frac{1}{2}\max\Big(\theta,1\Big) = \frac{1}{2}\theta, for\ \theta = 1,$$
$$\Leftrightarrow \theta \in \left\{0,1\right\},$$

$\forall\ \theta, \xi \in X$. Hence, $\{0\}$, $\{0, \frac{2}{3}\}$ and $\{0, 1\}$ are closed with respect to weak-partial b-metric space.

The geometrical representation of X defined by $\varrho_b : X \times X \to X$ is given in Table 1.

Table 1 The geometric representation of $d(\theta, \xi) \in X$

$\varrho_b : X \times X$	0	$\frac{1}{2}$	$\frac{2}{3}$	1
0	$(0, 0)$	$\left(0, \frac{1}{2}\right)$	$\left(0, \frac{2}{3}\right)$	$(0, 1)$
$\frac{1}{2}$	$\left(\frac{1}{2}, 0\right)$	$\left(\frac{1}{2}, \frac{1}{2}\right)$	$\left(\frac{1}{2}, \frac{2}{3}\right)$	$\left(\frac{1}{2}, 1\right)$
$\frac{2}{3}$	$\left(\frac{2}{3}, 0\right)$	$\left(\frac{2}{3}, \frac{1}{2}\right)$	$\left(\frac{2}{3}, \frac{2}{3}\right)$	$\left(\frac{2}{3}, 1\right)$
1	$(1, 0)$	$\left(1, \frac{1}{2}\right)$	$\left(1, \frac{2}{3}\right)$	$(1, 1)$

We claim that $\{0\}, \{\frac{2}{3}\}$ are not two fixed point of f and g. Now we show that for all $\theta, \xi \in X$ the inequality (17) is satisfied. To see this, consider the following cases.

Case 1 For $(\theta, \xi) \in \{0, \frac{2}{3}\}$ and using Definition 4, we have

$$\begin{aligned} \mathcal{H}(f\theta, g\xi) &= \mathcal{H}\left(\left[0, \frac{\theta}{2}\right], \left[0, \frac{\xi}{2}\right]\right) \\ &= \frac{1}{2}\left\{\sup\left(\left[0, \frac{\theta}{2}\right], \left[0, \frac{\xi}{2}\right]\right) + \sup\left(\left[0, \frac{\xi}{2}\right], \left[0, \frac{\theta}{2}\right]\right)\right\}. \end{aligned} \tag{44}$$

$$\begin{aligned} \sup\left(\left[0, \frac{\theta}{2}\right], \left[0, \frac{\xi}{2}\right]\right) &= \max\left\{\varrho_b\left(0, \left[0, \frac{\xi}{2}\right]\right), \varrho_b\left(\frac{\theta}{2}, \left[0, \frac{\xi}{2}\right]\right)\right\}, \\ &= \max\left\{0, \frac{\theta^2 + 2\theta}{8}\right\}, \\ &= \frac{\theta^2 + 2\theta}{8}. \end{aligned} \tag{45}$$

$$\begin{aligned} \sup\left(\left[0, \frac{\mu}{2}\right], \left[0, \frac{\theta}{2}\right]\right) &= \max\left\{\varrho_b\left(0, \left[0, \frac{\theta}{2}\right]\right), \varrho_b\left(\frac{\xi}{2}, \left[0, \frac{\kappa}{2}\right]\right)\right\}, \\ &= \max\left\{0, \frac{\xi^2 + 2\mu}{8}\right\}, \\ &= \frac{\xi^2 + 2\xi}{8}. \end{aligned} \tag{46}$$

Using (45) and (46) in (17) we get

$$\mathcal{H}(f\theta, g\xi) = \frac{\theta^2 + \xi^2 + 2\theta + 2\xi}{16}.$$

Next we calculate the following distances:

$$\begin{aligned} \varrho_b(\theta, \xi) &= \frac{\theta^2 + \xi^2 - 2\theta\xi + \xi}{2}, \\ \varrho_b(\theta, f\theta) &= \varrho_b\left(\theta, \left[0, \frac{\theta}{2}\right]\right) = \frac{\theta^2 + 2\theta}{8}, \\ \varrho_b(\xi, g\xi) &= \varrho_b\left(\xi, \left[0, \frac{\xi}{2}\right]\right) = \frac{9\xi^2 - 24\xi + 40}{72}, \\ \varrho_b(\theta, g\xi) &= \varrho_b\left(\theta, \left[0, \frac{\xi}{2}\right]\right) = \frac{\theta^2 + \theta}{2}, \end{aligned}$$

$$\varrho_b(\xi, f\theta) = \varrho_b\Big(y, \Big[0, \frac{\theta}{2}\Big]\Big) = \frac{\xi^2+\xi}{2},$$

$$\varrho_b(\theta, g\theta) = \varrho_b\Big(\theta, \Big[0, \frac{\theta}{2}\Big]\Big) = \frac{\theta^2+4\theta}{8},$$

$$\varrho_b(\xi, f\theta) = \varrho_b\Big(\xi, \Big[0, \frac{\xi}{2}\Big]\Big) = \frac{\xi^2+4\xi}{8}.$$

From (18) and (19), applying the above inequalities, it follows that

$$\mathcal{N}(\theta,\xi) \le \frac{\max\left\{\frac{\theta^2+\xi^2-2\theta\xi+\xi}{2}, \frac{\theta^2+2\theta}{8}+\frac{9\xi^2-24\mu+40}{72}, \frac{\theta^2+\theta}{2}+\frac{\xi^2+\xi}{2}+\tau\right\}}{\frac{\theta^2+2\theta}{8}+\frac{9\xi^2-24\xi+40}{72}+\gamma}, \tag{47}$$

$$\le \frac{\frac{\theta^2+\theta}{2}+\frac{\xi^2+y}{2}+\tau}{\frac{\theta^2+2\theta}{8}+\frac{9\xi^2-24\xi+40}{72}+\gamma}, \tag{48}$$

$$k = \mathcal{N}(\theta,\xi) := \frac{36(\theta^2+\xi^2+\theta+\xi+2\tau)}{9\theta^2+9\mu^2+18\theta-24\xi+40+72\gamma} \tag{49}$$

and

$$\mathcal{Q}(\theta,\xi) := \max\left\{\frac{\theta^2+\xi^2-2\theta\xi+\xi}{2}, \frac{\theta^2+4\theta}{8}, \frac{\xi^2+4\xi}{8}\right\},$$
$$= \frac{\theta^2+\xi^2-2\theta\xi+\xi}{2}. \tag{50}$$

Therefore, using (49), (50) in (17) reduces to

$$s^{\alpha}\frac{\theta^2+\xi^2+2\theta+2\xi}{16} \le \frac{36(\theta^2+\xi^2+\theta+\xi+2\tau)}{9\theta^2+9\xi^2+18\theta-24\xi+40+72\gamma} \times \left(\frac{\theta^2+\xi^2-2\theta\xi+\xi}{2}\right), \tag{51}$$
$$\le \frac{36(\theta^2+\xi^2+\theta+\xi+2\tau)(\theta^2+\xi^2-2\theta\xi+\xi)}{2s^{\alpha}(9\theta^2+9\xi^2+18\theta-24\xi+40+72\gamma)}.$$

Using (17) to determine the proper values of τ, γ, α, and $s \ge 1$ in X, we may conclude that all of the presumptions of Theorem 6 are met.

Case 2 For $(\theta, \xi) \in \left\{\frac{2}{3}, 1\right\}$ and using Definition 4, we obtain

$$\mathcal{H}(f\theta, g\xi) = \mathcal{H}\Big(\Big[0, \frac{2-\theta}{2}\Big], \Big[0, \frac{\xi}{2}\Big]\Big),$$

$$= \frac{1}{2}\left\{\sup\left(\left[0, \frac{2-\theta}{2}\right], \left[0, \frac{\xi}{2}\right]\right) + \sup\left(\left[0, \frac{\xi}{2}\right], \left[0, \frac{2-\theta}{2}\right]\right)\right\}. \tag{52}$$

$$\begin{aligned}\sup\left(\left[0, \frac{2-\theta}{2}\right], \left[0, \frac{\xi}{2}\right]\right) &= \max\left\{\varrho_b\left(0, \left[0, \frac{\xi}{2}\right]\right), \varrho_b\left(\frac{2-\theta}{2}, \left[0, \frac{\xi}{2}\right]\right)\right\}, \\ &= \max\left\{0, \frac{\theta^2 + \xi^2 - 6\theta + 4\xi - 2\theta\xi + 8}{8}\right\}, \\ &= \frac{\theta^2 + \xi^2 - 6\theta + 4\xi - 2\theta\xi + 8}{8}. \end{aligned} \tag{53}$$

$$\begin{aligned}\sup\left(\left[0, \frac{\xi}{2}\right], \left[0, \frac{2-\theta}{2}\right]\right) &= \max\left\{\varrho_b\left(0, \left[0, \frac{2-\theta}{2}\right]\right),\right. \\ &\qquad \left.\varrho_b\left(\frac{\xi}{2}, \left[0, \frac{2-\theta}{2}\right]\right)\right\}, \\ &= \max\left\{0, \frac{\theta^2 + \xi^2 - 6\theta - 4\xi + 2\theta\xi + 8}{8}\right\}, \\ &= \frac{\theta^2 + \xi^2 - 6\theta - 4\xi + 2\theta\xi + 8}{8}. \end{aligned} \tag{54}$$

Using (53) and (54) in (52) we get

$$\begin{aligned}\mathcal{H}(f\theta, g\xi) &= \frac{1}{2}\left\{\frac{\theta^2 + \xi^2 - 6\theta - 4\xi + 2\theta\xi + 8}{8} + \right. \\ &\qquad \left.\frac{\theta^2 + \xi^2 - 6\theta - 4\xi + 2\theta\xi + 8}{8}\right\}, \\ &= \frac{\theta^2 + \xi^2 - 6\theta - 4\xi + 2\theta\xi + 8}{8}. \end{aligned}$$

The same way, we calculate the following metrics:

$$\begin{aligned}\varrho_b(\theta, \xi) &= \frac{1}{2}\left(\theta - \xi\right)^2 + \frac{1}{2}\max\left(\theta, \xi\right), \\ &= \frac{\theta^2 + \xi^2 + \theta - 2\theta\xi}{2}, \\ \varrho_b(\theta, f\theta) &= \varrho_b\left(\theta, \left[0, \frac{2-\theta}{2}\right]\right) = \min\left\{\varrho_b\left(\theta, 0\right), \varrho_b\left(\theta, \frac{2-\theta}{2}\right)\right\}, \end{aligned}$$

$$= \frac{9\theta^2 - 8\theta + 4}{8}.$$

$$\varrho_b(\xi, g\xi) = \varrho_b\left(\xi, \left[0, \frac{\xi}{2}\right]\right) = \min\left\{\varrho_b\left(\xi, 0\right), \varrho_b\left(\xi, \frac{\xi}{2}\right)\right\},$$

$$= \min\left\{\frac{\xi^2 + 1}{2}, \frac{\xi^2 + 4\xi}{8}\right\} = \frac{\xi^2 + 4\xi}{8}.$$

$$\varrho_b(\theta, g\xi) = \varrho_b\left(\theta, \left[0, \frac{\xi}{2}\right]\right) = \min\left\{\varrho_b\left(\theta, 0\right), \varrho_b\left(\theta, \frac{\xi}{2}\right)\right\},$$

$$= \min\left\{\frac{\theta^2 + \theta}{2}, \frac{4\theta^2 + \xi^2 - 4\theta\xi + 4\theta}{8}\right\},$$

$$= \frac{4\theta^2 + \xi^2 - 4\theta\xi + 4\theta}{8}.$$

$$\varrho_b(\xi, f\theta) = \varrho_b\left(\xi, \left[0, \frac{2-\theta}{2}\right]\right) = \min\left\{\varrho_b\left(\xi, 0\right), \varrho_b\left(\xi, \frac{2-\theta}{2}\right)\right\},$$

$$= \frac{4\xi^2 + \theta^2 - 4\theta - 4\xi + 4\theta\xi + 4}{8}.$$

$$\varrho_b(\theta, g\theta) = \varrho_b\left(\theta, \left[0, \frac{\theta}{2}\right]\right) = \min\left\{\varrho_b(\theta, 0), \varrho_b(\theta, \frac{\theta}{2})\right\},$$

$$= \frac{\theta^2 + 4\theta}{8}.$$

$$\varrho_b(\xi, f\xi) = \varrho_b\left(\xi, \left[0, \frac{2-\xi}{2}\right]\right) = \min\left\{\varrho_b\left(\xi, 0\right)\varrho_b\left(\xi, \frac{2-\xi}{2}\right)\right\},$$

$$= \frac{9\xi^2 - 8\xi + 4}{8}.$$

Using (18) and (19) with the above inequalities, it follows that

$$k = \mathcal{N}\left(\theta, \xi\right) := \frac{\max\left\{\begin{array}{c}\frac{\theta^2+\xi^2+\theta-2\theta\mu}{2}, \frac{9\theta^2-8\theta+4}{8} + \frac{\xi^2+4\mu}{8}, \\ \frac{4\theta^2+\xi^2-4\theta\xi+4\theta}{8} + \frac{4^2+\theta^2-4\theta-4\xi+4\theta\xi+4}{8} + \tau\end{array}\right\}}{\frac{9\theta^2-8\theta+4}{8} + \frac{\xi^2+4\xi}{8} + \gamma},$$

$$\leq \frac{\max\left\{\frac{\theta^2+\xi^2+\theta-2\theta\xi}{2}, \frac{9\theta^2+\xi^2-8\theta+4\xi+4}{8}, \frac{5\theta^2+5\xi^2-4\xi+4}{8} + \tau\right\}}{\frac{9\theta^2+\xi^2-8\theta+4\xi+4}{8} + \frac{\xi^2+4\xi}{8} + \gamma},$$

$$= \frac{\frac{5\theta^2+5\xi^2-4\xi+4}{8} + \tau}{\frac{9\theta^2+\xi^2-8\theta+4\xi+4}{8} + \gamma},$$

$$= \frac{5\theta^2 + 5\xi^2 - 4\xi + 4 + 8\tau}{9\theta^2 + \xi^2 - 8\theta + 4\xi + 4 + 8\gamma}, \tag{55}$$

and

$$\begin{aligned} \mathcal{Q}\Big(\theta, \xi\Big) :&= \max\left\{\frac{\theta^2 + \xi^2 + \theta - 2\theta\mu}{2}, \frac{\theta^2 + 4\theta}{8}, \frac{9\xi^2 - 8\xi + 4}{8}\right\}, \\ &= \frac{9\xi^2 - 8\xi + 4}{8}. \end{aligned} \tag{56}$$

Therefore, using (55), (56) in (17) reduces to

$$\begin{aligned} \frac{\theta^2 + \xi^2 - 6\theta - 4\xi + 2\theta\xi + 8}{8} &\leq \frac{5\theta^2 + 5\xi^2 - 4\xi + 4 + 8\tau}{9\theta^2 + \xi^2 - 8\theta + 4\xi + 4 + 8\gamma} \times \\ &\quad \frac{9\xi^2 - 8\xi + 4}{8}, \\ &\leq \frac{(5\theta^2 + 5\xi^2 - 4\xi + 4 + 8\tau)(9\xi^2 - 8\xi + 4)}{8s^\alpha(9\theta^2 + \xi^2 - 8\theta + 4\xi + 4 + 8\gamma)}. \end{aligned}$$

Similarly, for case 2, we use (17) to determine the proper values of τ, γ, α, and $s \geq 1$ in X. Based on this, we conclude that all of the assumptions of Theorem 6 are met.

Case 3 For $\theta \in \left\{0, \frac{2}{3}\right\}$, $\xi \in \left\{\frac{2}{3}, 1\right\}$ and using Definition 4, we obtain

$$\begin{aligned} \mathcal{H}(f\theta, g\xi) &= \mathcal{H}\left(\left[0, \frac{2-\theta}{2}\right], \left[0, \frac{2\xi+1}{2}\right]\right), \\ &= \frac{1}{2}\left\{\sup\left(\left[0, \frac{2-\theta}{2}\right], \left[0, \frac{2\xi+1}{2}\right]\right) + \right. \\ &\quad \left. \sup\left(\left[0, \frac{2\xi+1}{2}\right], \left[0, \frac{2-\theta}{2}\right]\right)\right\}. \end{aligned} \tag{57}$$

$$\begin{aligned} \sup\left(\left[0, \frac{2-\theta}{2}\right], \left[0, \frac{2\xi+1}{2}\right]\right) &= \max\left\{\varrho_b\left(0, \left[0, \frac{2\xi+1}{2}\right]\right),\right. \\ &\quad \left. \varrho_b\left(\frac{2-\theta}{2}, \left[0, \frac{2\xi+1}{2}\right]\right)\right\}, \\ &= \max\left\{0, \frac{\theta^2 - 6\theta + 8}{8}\right\}, \\ &= \frac{\theta^2 - 6\theta + 8}{8}. \end{aligned} \tag{58}$$

$$\begin{aligned}
\sup\left(\left[0, \frac{2\xi+1}{2}\right], \left[0, \frac{2-\theta}{2}\right]\right) &= \max\left\{\varrho_b\left(0, \left[0, \frac{2-\theta}{2}\right]\right),\right. \\
&\qquad \left.\varrho_b\left(\frac{2\xi+1}{2}, \left[0, \frac{2-\theta}{2}\right]\right)\right\}, \\
&= \max\left\{0, \frac{4\xi^2+\theta^2+4\mu-2\theta+1}{8}\right\}, \\
&= \frac{4\xi^2+\theta^2+4\xi-2\theta+1}{8}. \qquad (59)
\end{aligned}$$

Using (58) and (59) in (57) we get

$$\begin{aligned}
\mathcal{H}(f\theta, g\xi) &= \frac{1}{2}\left\{\frac{\theta^2-6\theta+8}{8} + \frac{4\xi^2+\theta^2+4\xi-2\theta+1}{8}\right\}, \\
&= \frac{2\theta^2+4\xi^2-8\theta+4\xi+9}{16}.
\end{aligned}$$

Similarly, we generate the following metrics:

$$\begin{aligned}
\varrho_b(\theta, \xi) &= \frac{1}{2}\left(\theta-\xi\right)^2 + \frac{1}{2}\max\left(\theta, \xi\right) = \frac{\theta^2+\xi^2-2\theta\xi+\xi}{2}, \\
\varrho_b(\theta, f\theta) &= \varrho_b\left(\theta, \left[0, \frac{\theta}{2}\right]\right) = \frac{\theta^2+4\theta}{8}, \\
\varrho_b(\xi, g\xi) &= \varrho_b\left(\xi, \left[0, \frac{\xi}{2}\right]\right) = \frac{\xi^2+4\xi}{8}, \\
\varrho_b(\theta, g\xi) &= \varrho_b\left(\theta, \left[0, \frac{\xi}{2}\right]\right) = \frac{4\theta^2+\xi^2-4\theta\xi+4\theta}{8}, \\
\varrho_b(\xi, f\theta) &= \varrho_b\left(\xi, \left[0, \frac{\theta}{2}\right]\right) = \frac{\theta^2+\xi^2+4\xi-2\theta\xi}{8}, \\
\varrho_b(\theta, g\theta) &= \varrho_b\left(\theta, \left[0, \frac{2\theta+1}{2}\right]\right) = \frac{\theta^2+\theta}{2}, \\
\varrho_b(\xi, f\xi) &= \varrho_b\left(\xi, \left[0, \frac{2-\xi}{2}\right]\right) = \frac{9\theta^2-8\xi+4}{8}.
\end{aligned}$$

Using (18) and (19) with the above metrics yields to

$$k = \mathcal{N}\left(\theta, \xi\right) := \frac{\max\left\{\begin{matrix}\frac{\theta^2+\xi^2-2\theta\xi+\xi}{2}, \frac{\theta^2+4\theta}{8} + \frac{\xi^2+4\xi}{8}, \\ \frac{4\theta^2+y^2-4\theta\xi+4\theta}{8} + \frac{\theta^2+\xi^2-2\theta\xi+4\xi}{8} + \tau\end{matrix}\right\}}{\frac{\theta^2+4\theta}{8} + \frac{\xi^2+4\xi}{8} + \gamma},$$

$$\leq \frac{\max\left\{\frac{\theta^2+\xi^2-2\theta\xi+\xi}{2}, \frac{\theta^2+\xi^2+4\theta+4\xi}{8}, \frac{5\theta^2+2\xi^2+4\theta+4\xi-6\theta\xi}{8}+\tau\right\}}{\frac{\theta^2+\xi^2+4\theta+4\xi}{8}+\gamma},$$

$$\leq \frac{5\theta^2+2\xi^2+4\theta+4\xi-6\theta\xi+8\tau}{\theta^2+\mu^2+4\theta+4\xi+8\gamma}.$$

and

$$\begin{aligned} \mathcal{Q}\Big(\theta,\xi\Big) &:= \max\left\{\frac{\theta^2+\xi^2-2\theta\xi+\xi}{2}, \frac{\theta^2+\theta}{2}, \frac{9\xi^2-8\xi+4}{8}\right\}, \\ &= \frac{\theta^2+\theta}{2}. \end{aligned} \tag{60}$$

Therefore, using (60), (61) in (17) reduces to

$$\frac{2\theta^2+4\xi^2-8\theta+4\xi+9}{16} \leq \frac{(5\theta^2+2\xi^2+4\theta+4\xi-6\theta\xi+8\tau)(\theta^2+\theta)}{s^{\alpha}(2\theta^2+2\xi^2+8\theta+8\xi+16\gamma)}.$$

Using (17) to determine the proper values of τ, γ, α, and $s \geq 1$ in X, we may conclude that all of the presumptions of Theorem 6 are met. It is noted that there are multiple fixed points in X for a multivalued mapping f, g. As a result, f, g at least share a fixed point. As a result, f, g has two fixed points: 0 and $\frac{2}{3}$. To support this, have a look at the inequality

$$\begin{aligned} \varrho_b(J, J^*) &\geq \frac{\gamma s^{2\alpha}-\tau}{2} \\ \varrho_b\left(0, \frac{2}{3}\right) &\geq \frac{\gamma s^{2\alpha}-\tau}{2} \\ \varrho_b\left(0, \frac{2}{3}\right) = \frac{1}{2}\left(0-\frac{2}{3}\right)^2 + \frac{1}{2}\max\left(0, \frac{2}{3}\right) &\geq \frac{\gamma s^{2\alpha}-\tau}{2} \\ \frac{5}{9} &\geq \frac{\gamma s^{2\alpha}-\tau}{2}, \end{aligned} \tag{61}$$

if we choose $\alpha = \frac{1}{4}$, $\tau, \gamma = 1$ and $s = 2$, the above inequality reduces to

$$\begin{aligned} \frac{5}{9} &\geq \frac{2^{\frac{1}{2}}-1}{2}, \\ \frac{5}{9}. &\geq \frac{\sqrt{2}-1}{2}, \end{aligned}$$

which contradicts itself. This demonstrates that every condition stated in Theorem 6 is met. Thus, two fixed points of f and g are $\{0\}$ and $\{\frac{2}{3}\}$.

The following corollary is motivated from [11] [Theorem 16]. We prove the results in multivalued weak-partial b-metric space.

Corollary 1 *Given a complete weak-partial b-metric space (X, ϱ_b), let X be a paracompact space of X. Consider a multivalued mapping $g : X \to P_{cl}^{\varrho_b}(X)$, such that $g\theta \neq 0$ for each $\theta, \xi \in X$, $q, r \geq 2$, and $H(g\theta, g\xi) \geq 0$, we have*

(i) There exists $k \in (0, 1)$ such that, for each $\theta, \xi \in X$,

$$s^\alpha sH(g\theta, g\xi) \leq k\mathcal{M}(\theta, \xi) \tag{62}$$

where

$$k = \mathcal{N}(\theta, \xi) := \frac{\varrho_b(\theta, g\xi) + \varrho_b(\xi, g\theta) + \tau}{\varrho_b(\theta, g\theta) + \varrho_b(\xi, g\xi) + \gamma}, \tag{63}$$

and

$$\mathcal{M}(\theta, \xi) = \max\left\{\varrho_b(\theta, \xi), \frac{\varrho_b(\theta, g\theta) + \varrho_b(\xi, g\xi)}{q}, \frac{\varrho_b(\theta, g\xi) + \varrho_b(\xi, g\theta)}{sr}\right\},$$

for all $\theta, \xi \in X$, $s \geq 1$.

(ii) $\exists\, \jmath \in g\xi$ such that, for any $\theta \in X$, $\xi \in g\theta$, and $\epsilon > 0$

$$\varrho_b(\xi, \jmath) \leq \mathcal{H}(g\theta, g\xi) + \epsilon. \tag{64}$$

Additionally, presumptively the following holds true:

(a) At least one fixed point $\jmath \in X$ exists for g.
(b) For every $\theta \in X$, $\{g^{\frac{n}{2}}\theta\}$ converges to a fixed point for n.
(c) $\varrho_b(\jmath, \jmath^*) \geq \frac{\gamma s^{2\alpha} - \tau}{2}$ is the result if $g\jmath = \{\jmath\}$ and $\jmath, \jmath^*$ are distinct tight fixed points of g satisfying a g_1 space in X. X is a paracompact space.

Proof Suppose that g has no fixed points then $\varrho_b(\theta_0, g\theta_0) > 0$ for all $\theta \in X$. We start by constructing a sequence $\{\theta_n\}$, respectively, in the following way.

Let θ_n be an arbitrary point in X, and X is a paracompact space as in Definition 8 and $\theta_0 \in g\theta_0$ such that $\varrho_b(\theta_0, \theta_1) \leq \varrho_b(\theta_0, g\theta_0)$. If $\theta_0 = \theta_1$ or $\theta_1 = g\theta_1$ we deduce that g has at least one fixed point θ_1 in X. Thus, the proof is completed. On contrary to that, let $\theta_0 \neq \theta_1$, $\theta_1 \neq g\theta_1$, and $\varrho_b(\theta_1, g\theta_1) > 0$, which implies that $\mathcal{H}(g\theta_0, g\theta_1) > 0$.

From Eq. (20), we choose $\theta_2 \in g\theta_1$ such that

$$\varrho_b(\theta_1, \theta_2) \leq \mathcal{H}(g\theta_0, g\theta_1) + \epsilon. \tag{65}$$

Suppose that $\epsilon = \left(\frac{1}{\sqrt{k}} - 1\right)\mathcal{H}(g\theta_0, g\theta_1)$ in (66), we get

$$\begin{aligned}
\varrho_b(\theta_1, \theta_0) &\leq \mathcal{H}(g\theta_0, g\theta_1) + \left(\frac{1}{\sqrt{k}} - 1\right)\mathcal{H}(g\theta_0, g\theta_1), \\
&\leq \mathcal{H}(g\theta_0, g\theta_1) + \frac{1}{\sqrt{k}}\mathcal{H}(g\theta_0, g\theta_1) - \mathcal{H}(g\theta_0, g\theta_1), \\
&\leq \frac{1}{\sqrt{k}}\mathcal{H}(g\theta_0, g\theta_1), \forall\, 0 < k < 1.
\end{aligned} \tag{66}$$

For $\theta_2 \in g\theta_1$ there exists $\theta_3 \in g\theta_2$ such that

$$\varrho_b(\theta_2, \theta_3) \leq \frac{1}{\sqrt{k}}\mathcal{H}(g\theta_1, g\theta_2), \forall\, 0 < k < 1.$$

Continuing this process, we obtain a sequence $\{\theta_n\}$ in X such that

$$\begin{aligned}
\varrho_b(\theta_n, \theta_{n+1}) &\leq \frac{1}{\sqrt{k}}\mathcal{H}(g\theta_{n-1}, g\theta_n), \forall\, 0 < k < 1. \\
&\leq \frac{1}{s^\alpha}\frac{1}{\sqrt{k}}k\mathcal{M}(\theta_{n-1}, \theta_n), \\
&\leq \frac{\sqrt{k}}{s^\alpha}\mathcal{M}(\theta_{n-1}, \theta_n).
\end{aligned} \tag{67}$$

From (68), using (62), (63), and (64) by mathematical induction, we obtain

$$\varrho_b(\theta_n, \theta_{n+1}) \leq \frac{(\sqrt{s\beta_n})^n}{s^\alpha}\varrho_b(\theta_{n-1}, \theta_n), \tag{68}$$

where $\beta_n = \frac{(\varrho_b(\theta_{n-1},\theta_n)+\varrho_b(\theta_n,\theta_{n+1}))+\tau}{\varrho_b(\theta_{n-1},\theta_n)+\varrho_b(\theta_n,\theta_{n+1}))+\gamma}$. Let $\sigma = \frac{(\sqrt{s\beta_n})^n}{s^\alpha}$, by Lemmas 3 and 4, we have

$$\varrho_b(\theta_n, \theta_{n+1}) \leq \sigma\varrho_b(\theta_{n-1}, \theta_n), \tag{69}$$

which shows that $\{\theta_n\}$ is a Cauchy sequence. The remaining stages adhere to the comparable demonstration of Theorem 6. This concludes the proof.

4 An Application to NFDE

In order to obtain a common solution for a three-point boundary value nonlinear fractional differential equation (NFDE), this section applies Theorem 6. In this way, we can utilize Rhoades [30] multivalued-type mappings in weak-partial b-metric spaces. In order to do this, we use the Caputo fractional differential equation, which has numerous mathematical applications, including probability theory, acoustics, digital data processing, image processing, electrical signals, and physics (see [36]).

We investigate the Caputo derivative of the nonlinear fractional differential equation with the fractional order.

A continuous function $f : [0, \infty) \to \mathbf{R}$ has the following form for its fractional derivative:

$$({}^{C}D_{t}^{\beta})f(t) = \frac{1}{\Gamma(n-\beta)} \int_{a}^{t} (t-\gamma)^{n-\beta-1} K^{n}(\gamma)d\gamma, \; (n-1 < \beta, n = [\beta]+1), \tag{70}$$

where the integer component of the real number β is indicated by $[\beta]$ (see [5]). Furthermore, the order α fractional integral of Riemann–Liouville is given by

$$(I_{\gamma}^{\beta})f(t) = \frac{1}{\Gamma(\beta)-1} \int_{0}^{t} (t-\gamma)^{\beta} K(\gamma)d\gamma \; (\beta > 0). \tag{71}$$

Inspired by Ahmad and Nieto [1], we have the following nonlinear three-point boundary valued problem of sequential fractional differential equation.

$$\begin{cases} {}^{C}D^{\beta}(D+\zeta)\theta(t) = K(t, \theta(t)), \; t \in (0,1), 1 < \alpha \leq 2, \\ \theta(0) = 0, \theta'(0) = 0, \theta(1) = \tau\theta(\delta), \; 0 < \delta < 1, \end{cases} \tag{72}$$

where $K : [0, 1] \times \mathbf{R} \to \mathbf{R}$ is a continuous function, θ is a positive real integer, and β is a real number such that θ is the Caputo fractional derivative of order α. D is the ordinary derivative.

$$\tau \neq \frac{\zeta + e^{-\zeta} - 1}{\zeta\delta + e^{-\zeta\delta} - 1}.$$

Let $X = C(I)(I = [0, 1])$ represent the space containing the continuous function given on I. Define a Banach space $(X, \|.\|)$ with $I := [0, T], T > 0$. Let $C(I, X)$ be the Banach space for every continuous function from I into X with the norm $\|\theta\| := \sup|\theta(t)| = \psi, t \in I$ for $\theta \in C(I, X)$. Examine a collection of continuous real-valued functions that are defined in the function space $C(X, \mathbf{R})$ and on X. A weak-partial b-metric norm $\|.\|$ on X can be defined as follows:

$$\varrho_b(\theta, \xi) = \sup_{t \in I} |\theta(t) - \xi(t)|^p + \alpha, \tag{73}$$

$\forall \theta, \xi \in C(I, X), p > 1$ and $\alpha > 0$. This space is a weak-partial b-metric in its entirety. Subject to the boundary constraints, the nonlinear fractional equation (72) can be solved as

$$\theta(t) = \int_{0}^{t} e^{-\delta(t-\gamma)} \left(\frac{1}{\Gamma(\beta)} \int_{0}^{\gamma} (t-\gamma)^{\beta-1} f(r, \theta(r))dr \right) ds +$$

$$A(t)\Big[\tau\int_0^{\delta} e^{-\zeta(\delta-\gamma)}\Big(\frac{1}{\Gamma(\beta)}\int_0^{\gamma}(\gamma-r)^{\beta-1}K(r,\theta(r))dr\Big)d\gamma -$$
$$\int_0^{1} e^{-\zeta(t-\gamma)}\Big(\frac{1}{\Gamma(\beta)}\int_0^{\gamma}(\gamma-r)^{\beta-1}K(r,\theta(r))dr\Big)d\gamma\Big], \tag{74}$$

where $A(t)=\frac{1}{\nabla}\Big(\zeta t+e^{-\zeta t}-1\Big)$ and $\nabla=\zeta+e^{-\zeta}-1-\tau\Big(\zeta\delta+e^{-\zeta\delta}-1\Big)$.

A fractional differential integral equation (74) can be solved by a function $\theta\in C(I,X)$ if and only if the nonlinear fractional differential equation (72) has a solution at θ.

Inspired by the conversation above, we establish the theorem below.

Theorem 7 *Assume the following scenario is true:*

(i) $K\in C(I\times X, P_{cl}^{\varrho_b}(X))$ is sequential continuous;
(iii) there exists a continuous function $K:[0,1]\times\mathbf{R}\to\mathbf{R}_{+}$, such that

$$|K(t,\theta(\gamma))-K(t,\xi(\gamma))|\le s^{-\alpha}\psi(t)\|\theta(\gamma)-\xi(\gamma)\|, \tag{75}$$

for all $t\in[0,1]$ and for all $\theta,\xi\in X$, where $0<\psi(t)<t$ for $t>0$, such that

$$s^{\alpha}\mathcal{H}(f\theta\backslash\{\theta\},g\xi\backslash\{\xi\})\le\varrho_b(f\theta,g\xi)\le k\mathcal{Q}(\theta,\xi),$$

where

$$k=\mathcal{N}(\theta,\xi):=\frac{\max\Big\{\varrho_b(\theta,\xi),\varrho_b(\theta,f\theta)+\varrho_b(\xi,g\xi),\varrho_b(\theta,g\xi)+\varrho_b(\xi,f\theta)+\tau\Big\}}{\varrho_b(\theta,f\theta)+\varrho_b(\xi,g\xi)+\gamma},$$

and

$$\mathcal{Q}(\theta,\xi):=\max\Big\{\varrho_b(\theta,\xi),\varrho_b(\theta,g\theta),\varrho_b(\xi,f\xi)\Big\}.$$

Also $\psi(t)\le 1$, that is,

$$\psi(t)=\Big(t^{p-1}\Big)\Big(\frac{t^{-\beta p}}{\beta(1+\theta p)\Gamma(\beta)}+|A(t)|^{p}\Big[\frac{\tau^{p}\eta^{\beta p}-e^{-\theta p(t-1)}}{\beta(1+\theta p)\Gamma(\beta)}\Big]\Big),$$

where $A(t)=\frac{1}{\nabla}\Big(\zeta t+e^{-\zeta t}-1\Big)$ and $\nabla=\zeta+e^{-\zeta}-1-\tau\Big(\zeta\eta+e^{-\zeta\delta}-1\Big)$.

Consequently, there is at least one common solution $\theta\in C(I,X)$ for the fractional differential equation (72).

Proof: Let us define $f\,g:C(I)\to C(I)$ by

$$
\begin{aligned}
fg\theta(t) &= \int_0^t e^{-\zeta(t-\gamma)}\left(\frac{1}{\Gamma(\beta)}\int_0^s(\gamma-r)^{\beta-1}K(r,\theta(r))dr\right)d\gamma + \\
& A(t)\left[\beta\int_0^\delta e^{-\zeta(\delta-\gamma)}\left(\frac{1}{\Gamma(\beta)}\int_0^\gamma(\gamma-r)^{\beta-1}K(r,\theta(r))dr\right)d\gamma - \right.\\
& \left.\int_0^1 e^{-\zeta(t-\gamma)}\left(\frac{1}{\Gamma(\beta)}\int_0^\gamma(\gamma-r)^{\beta-1}K(r,\theta(r))dr\right)d\gamma\right], \qquad (76)
\end{aligned}
$$

for $t, \gamma \in [0, 1]$, then fg is sequential continuous. Assume that

$$
g\theta(t) = \frac{1}{\Gamma(\beta)}\int_0^\gamma(\gamma-r)^{\beta-1}K(r,\theta(r))dr. \qquad (77)
$$

We assert that fg is sequential continuous and contraction in order to demonstrate the existence of at least one common fixed point. Given any $\theta, \xi \in [0, T]$, we may demonstrate that fg is sequential continuous and contraction by letting $fg\theta \neq fg\xi$. Given (ii), we obtain

$$
\begin{aligned}
|fg\theta - fg\xi| &= \left|\left[\int_0^t e^{-\zeta(t-\gamma)}\left(\frac{1}{\Gamma(\beta)}\int_0^\gamma(t-\gamma)^{\beta-1}K(r,\theta(r))dr\right)d\gamma + \right.\right.\\
& A(t)\left[\beta\int_0^\delta e^{-\zeta(\delta-\gamma)}\left(\frac{1}{\Gamma(\beta)}\int_0^\gamma(\gamma-r)^{\beta-1}K(r,\theta(r))dr\right)d\gamma - \right.\\
& \left.\left.\int_0^1 e^{-\zeta(t-\gamma)}\left(\frac{1}{\Gamma(\beta)}\int_0^\gamma(\gamma-r)^{\beta-1}K(r,\theta(r))dr\right)d\gamma\right]\right] - \\
& \left[\int_0^t e^{-\zeta(t-\gamma)}\left(\frac{1}{\Gamma(\beta)}\int_0^\gamma(\gamma-r)^{\beta-1}K(r,\xi(r))dr\right)d\gamma + \right.\\
& A(t)\left[\tau\int_0^\delta e^{-\zeta(\delta-\gamma)}\left(\frac{1}{\Gamma(\beta)}\int_0^\gamma(\gamma-r)^{\beta-1}K(r,\xi(r))dr\right)d\gamma - \right.\\
& \left.\left.\left.\int_0^1 e^{-\zeta(t-\gamma)}\left(\frac{1}{\Gamma(\beta)}\int_0^\gamma(\gamma-r)^{\alpha-1}K(r,\xi(r))dr\right)d\gamma\right]\right]\right|^p + \beta
\end{aligned}
$$

$$
\begin{aligned}
&\leq \left|\int_0^t e^{-\sigma(t-s)}\left(\frac{1}{\Gamma(\alpha)}\int_0^s(s-u)^{\alpha-1}|K(u,\theta(u)) - K(u,\xi(u))|du\right)ds + \right.\\
& A(t)\left[\beta\int_0^\delta e^{-\zeta(\delta-\gamma)}\left(\frac{1}{\Gamma(\beta)}\int_0^\gamma(\gamma-r)^{\beta-1}|K(r,\theta(r)) - f(r,\xi(r))|dr\right)d\gamma\right.\\
& \left.\left.- \int_0^1 e^{-\zeta(t-\gamma)}\left(\frac{1}{\Gamma(\beta)}\int_0^\gamma(\gamma-r)^{\beta-1}|K(r,\theta(r)) - K(r,\xi(r))|dr\right)d\gamma\right]\right|^p + \beta \\
&\leq s^{-\alpha}\left[\int_0^t\left(d\gamma\right)^{\frac{1}{q}}\left(\int_0^t e^{-\zeta p(t-\gamma)}\left(\frac{\gamma^{\beta p}}{\beta\Gamma(\beta)}|\theta-\xi|^p\right)d\gamma\right.\right.
\end{aligned}
$$

$$+|A(t)|^p\Bigg[\tau^p\int_0^\delta e^{-\zeta p(\delta-\gamma)}\Big(\frac{\gamma^{\beta p}}{\beta\Gamma(\beta)}|\theta(\gamma)-\xi(\gamma)|^p\Big)d\gamma$$

$$-\int_0^1 e^{-\zeta p(t-\gamma)}\Big(\frac{\gamma^{\beta p}}{\beta\Gamma(\beta)}|\theta(\gamma)-\xi(\gamma)|^p\Big)d\gamma\Bigg]\Bigg)^{\frac{1}{p}}\Bigg]^p+\beta,$$

$$\leq s^{-\alpha}\int_0^t\Big(d\gamma\Big)^{\frac{p}{q}}\Big(\int_0^t e^{-\zeta p(t-\gamma)}\Big(\frac{\gamma^{\beta p}}{\beta\Gamma(\beta)}|\theta(\gamma)-\xi(\gamma)|^p\Big)d\gamma$$

$$+|A(t)|^p\Bigg[\tau^p\int_0^\delta e^{-\zeta p(\delta-\gamma)}\Big(\frac{\gamma^{\beta p}}{\beta\Gamma(\beta)}|\theta(\gamma)-\xi(\gamma)|^p\Big)ds$$

$$-\int_0^1 e^{-\zeta p(t-\gamma)}\Big(\frac{s^{\beta p}}{\beta\Gamma(\beta)}|\theta(\gamma)-\xi(\gamma)|^p\Big)d\gamma\Bigg]\Big)+\beta,$$

$$\leq s^{-\alpha}\Big(t\Big)^{p-1}\Big(\frac{t^{-\beta p}}{\beta(1+\zeta p)\Gamma(\alpha)}+|A(t)|^p\Big[\tau^p\frac{\delta^{\beta p}}{\beta(1+\zeta p)\Gamma(\beta)}$$

$$-\frac{e^{-\zeta p(t-1)}}{\beta(1+\theta p)\Gamma(\beta)}\Big]\Big)|\theta(\gamma)-\xi(\gamma)|^p+\alpha,$$

$$\leq \gamma^{-\beta}\Big(t^{p-1}\Big)\Big(\frac{t^{-\beta p}}{\beta(1+\zeta p)\Gamma(\beta)}+|A(t)|^p\Big[\frac{\tau^p\delta^{\beta p}-e^{-\zeta p(t-1)}}{\beta(1+\zeta p)\Gamma(\beta)}\Big]\Big)|\theta(\gamma)-\xi(\gamma)|^p+\beta,$$

$$\leq s^{-\alpha}\psi(t)|\theta(\gamma)-\xi(\gamma)|^p+\beta.$$

This implies that

$$|fg\theta-fg\xi|\ \leq s^{-\alpha}\psi(t)|\theta(\gamma)-\xi(\gamma)|^p+\beta.$$

Since $\psi(t)<1$. Thus for each $\theta,\xi\in X$, where X is a paracompact Hausdorff space, we have

$$\mathcal{H}(f\theta, g\xi)\ \leq s^{-\alpha}k\mathcal{Q}(\theta,\xi). \tag{78}$$

Hence, in contradiction to what we have claimed, there exists $\theta\in C(I)$, a shared fixed point of f and g. In other words, θ is a solution to the fractional nonlinear differential equation (72). As a result, f and g share at least one fixed point with X.

Declarations:

Funding: The authors declare that funding for this article is not available.

Competing interests/Conflict of interest: The authors declare that they have no competing interests.

Author's contributions: Each author made a substantial and equal contribution to the writing of this paper. The final manuscript was read and approved by all writers.

Availability of data and material: This study was not supported by any data.

Acknowledgements The reviewer's insightful remarks are much appreciated by the authors.

References

1. Ahmad, B., Nieto, J.J.: Sequential fractional differential equations with three-point boundary conditions. Comput. Math. Appl. **64**(10), 3046–3052 (2012)
2. Aliouche, A., Hamaizia, T.: Common fixed point theorems for multivalued mappings in b-metric spaces with an application to integral inclusions. J. Anal. 1–20 (2021)
3. Aydi, H., Abbas, M., Vetro, C.: Partial Hausdorff and Nadler's fixed point theorem on partial metric space. Topol. Appl. **159**(14), 3234–3242 (2012)
4. Bakhtin, I.A.: The contraction principle in quasi metric spaces. Funct. Anal. 26–37 (1989)
5. Baleanu, D., Rezapour, S., Mohammadi, H.: Some existence results on nonlinear fractional differential equations. Philos. Trans. R. Soc. Lond. Ser. A Math. Phys. Eng. Sci. **371**, 1–7 (1990)
6. Beg, I., Pathak, H.K.: A variant of Nadler's theorem on weak partial metric spaces with application to a homotopy result. Vietnam J. Math. **46**(3), 693–706 (2018)
7. Boriceanu, M., Bota, M., Petruşel, A.: Multivalued fractals in b-metric spaces. Central Eur. J. Math. **8**(2), 367–377 (2010)
8. Chifu, C., Petruşel, G.: Fixed point results for multi-valued Hardy-Rogers contractions in b-metric spaces. Filomat **31**(8), 2499–2507 (2017)
9. Czerwik, S.: Contraction mappings in b-metric spaces. Acta Math. Inform. Univ. Ostraviensis **1**, 5–11 (1993)
10. Czerwik, S.: Nonlinear set-valued contraction mappings in b-metric spaces. Atti Sem. Mat. Fis. Univ. Modena **46**, 263–276 (1998)
11. Demma, M., Vetro, P.: Picard sequence and fixed point results on b-metric spaces. J. Funct. Spaces. https://doi.org/10.1155/2015/189861
12. Dieu donné, J.: Une généralisation des espaces compacts. J. Math. Pures Appl. **23**, 65–76 (1944)
13. Dukic, D., Kadelburg, Z., Radenovic, S.: Fixed point of Geraghty-type mappings in various generalized metric spaces. Abstr. Appl. Anal. **2011**, Article ID 561245, 1–13. https://doi.org/10.1155/2011/561245
14. Heckmann, R.: Approximation of metric spaces by partial metric spaces. Appl. Categ. Struct. **7**(1), 71–83 (1999)
15. Jungck, G.: Compatible mappings and common fixed points. Int. J. Math. Math. Sci. **9**(4), 771–779 (1986)
16. Kaneko, H., Sessa, S.: Fixed point theorems for compatible multi-valued and single-valued mappings. Int. J. Math. Math. Sci. **12**, 257–262 (1989)
17. Kanwal, T., Hussain, A., Kumam, P., Savas, E.: Weak partial b-metric spaces and Nadler's Theorem. Mathematics **7**(4), 332 (2019)
18. Khojasteh, F., Abbas, M., Costache, S.: Two new types of fixed point theorems in complete metric spaces. Abstr. Appl. Anal. (2014)
19. Kirk, W., Shahzad, N.: Fixed point theory in distance spaces. Springer, Cham (2014)
20. Kubiak, T.: Fixed points for contractive type multi-functions. Math. Jpn. **30**, 89–101 (1985)
21. Kutbi, M.A., Karapınar, E., Ahmad, J., Azam, A.: Some fixed point results for multi-valued mappings in b-metric spaces. J. Inequal. Appl. **4**(1), 1–11 (2014)
22. Matthews, S.G.: Partial-metric topology. Ann. N. Y. Acad. Sci. **728**, 183–197 (1994)
23. Michael, E.: Continuous selections, I. Ann. Math. **63**(2), 361–382 (1956)
24. Nadler, S.B.: Multi-valued contraction mappings. Am. Pacific J. Math. **30**(2), 475–488 (1969)
25. Oltra, S., Valero, O.: Banach's fixed point theorem for partial metric spaces. Rend. Istid Mat Univer. Trieste **36**(1–2), 17–26 (2004)

26. Oltra, S., Romaguera, S., Sánchez-Pérez, E.A.: Bicompleting weightable quasi metric spaces and partial metric spaces. Rend. Ciric Mat palermo **51**(1), 151–162 (2002)
27. Paesano, D., Vetro, C.: Multi-valued F-contractions in 0-complete partial metric spaces with application to Volterra type integral equation. Revista de la Real Academia de Ciencias Exactas Fisicas Naturales **108**(2), 1005–1020 (2014)
28. Pathak, H.K.: Fixed point theorems for weak compatible multi-valued and single valued mappings. Acta Math. Hungar. **67**(1–2), 69–78 (1995)
29. Rhoades, B.E.: Two new fixed point theorems. General Math. Notes **27**(2), 123–132 (2015)
30. Rhoades, B.E.: A fixed point theorem for non-self set-valued mappings. Int. J. Math. Math. Sci. **20**(1), 9–12 (1977)
31. Roshan, J.R., Parvaneh, V., Kadelburg, Z.: Common fixed point theorems for weakly isotone increasing mappings in ordered b-metric spaces. J. Nonlinear Sci. Appl. **7**, 229–245 (2014)
32. Sessa, S.: On a weak commutativity condition of mappings in fixed point considerations. Publ. Inst. Math **32**(46), 149–153 (1982)
33. Shahzard, N., Valero, O.: On-0-complete partial metric spaces and quantitative fixed point technique. Abstr. Appl. Anal. **2013**, 1–11 (2013)
34. Shukla, S.: Partial b-metric spaces and fixed point theorems. Mediterr. J. Math. **11**(2), 703–711 (2014)
35. Stone, A.H.: Paracompactness and product spaces. Bull. Am. Math. Soc. **54**(10), 977–9822 (1948)
36. Zölzer, U. (ed.): DAFX: Digital Audio Effects, vol. 21, no. 2, pp. 48–49 (2002)
37. Wangwe, L., Kumar, S.: Common fixed point theorem for hybrid pair of mappings in a generalized (F, ξ, η)-contraction in weak partial b-metric spaces with an application. Adv. Theory Nonlinear Anal. Appl. **5**(4), 531–550 (2021)
38. Wangwe, L., Kumar, S.: Fixed point results for interpolative ψ-Hardy-Rogers type contraction mappings in quasi-partial b-metric space with some applications. J. Anal. 18 (2022). https://doi.org/10.1007/s41478-022-00456-4

Deep Learning Based Parametric Estimation in Double-Diffusive Convective Flow

Sumant Kumar, S. V. S. S. N. V. G. Krishna Murthy, and B. V. Rathish Kumar

Abstract The present research work employs the artificial neural network (ANN) based deep learning (DL) approach to analyze the parametric estimation in double-diffusive convective flow in an inverted T-shaped porous enclosure. The study develops and numerically simulates a Darcy-extended Brinkmann-Forchheimer mathematical model using the penalty finite element method. Minimal training data are gathered across various flow parameters for DL model training. The DL-based technique efficiently approximates convective heat and mass transport across unknown flow parameter sets, significantly reducing computational costs. Analysis reveals that higher Rayleigh numbers prominently trigger convective phenomena, offering insights into realistic impacts of other flow parameters within porous media.

Keywords ANN · Double-diffusion · Deep learning · FEM · Porous media

1 Introduction

Double-diffusive convection, driven by interacting thermal and solute gradients, is a ubiquitous phenomenon with far-reaching implications across scientific and industrial domains [1]. By modeling the complex interplay within porous structures, researchers gain invaluable insights into fluid flow dynamics, temperature distributions, and concentration patterns. This process underpins critical applications such as groundwater management, contaminant remediation, and enhanced oil recovery. Additionally, double-diffusion principles find utility in diverse fields, including biomedical engineering and solar energy systems [2–8].

S. Kumar · S. V. S. S. N. V. G. Krishna Murthy
School of Computer Engineering and Mathematical Sciences, Defence Institute of Advanced Technology, Pune, India
e-mail: skmurthy@diat.ac.in

B. V. R. Kumar (✉)
Department of Mathematics and Statistics, Indian Institute of Technology Kanpur, Kanpur, India
e-mail: bvrk@iitk.ac.in

R. K. Sharma et al. (eds.), *Mathematical Theory and Simulation of Scientific Problems*, Springer Proceedings in Mathematics & Statistics 487,
https://doi.org/10.1007/978-981-96-2579-6_10

Artificial neural networks (ANNs) have rapidly gained prominence across diverse scientific and industrial fields due to their exceptional capabilities. From autonomous vehicles to medical image analysis, ANNs have demonstrated remarkable success [9, 10]. Notably, their ability to approximate complex mathematical functions and estimate parameters has positioned them as powerful tools for modeling intricate systems. For instance, Mandal et al. [11] used ANNs to predict thermal behavior in a porous cavity, while Prince et al. [12, 13] employed them to estimate nanofluid formation even with limited data. Similarly, Filali et al. [14] leveraged ANNs for thermal behavior prediction in mixed convection. Kumar et al. [15] have employed the ANN for parametric estimation during the entropy generation study in a complex porous enclosure. These studies, alongside others [16–18], demonstrate the effectiveness of ANNs in predicting flow parameters within complex geometries. Trained ANNs can rapidly and accurately estimate unknown parameters using processed numerical data. This effectively translates to predicting heat and fluid flow behavior by feeding post-processed data from physics-informed simulations into the ANN model. The high level of agreement with established methods like the finite element method (FEM) highlights the potential of ANNs for approximating parameter influences in complex mathematical models.

The research work reported till on double-diffusive natural convection has primarily focused on conventional geometries. To address this gap, this study delves into the intricate dynamics of an inverted T-shaped porous enclosure, incorporating the complexities of chemical reactions, thermal radiation, and Soret/Dufour effects. A comprehensive mathematical model was developed to simulate these multi-physics interactions. Numerical simulations were conducted across a wide range of flow parameters. To enhance efficiency, an ANN-based deep learning model was trained to predict heat and mass transfer rates, significantly outperforming traditional numerical methods. This integrated approach provides valuable insights into the intricate behavior of double-diffusive convection within the inverted T-shaped enclosure, with potential applications in solar energy systems and thermal engineering.

2 Development of a Mathematical Model

Figure 1 presents an inverted T-shaped porous cavity filled with a hybrid nanofluid consisting of multi-walled carbon nanotubes (MWCNTs) and ferric oxide (Fe_3O_4) nanoparticles suspended in water at a fixed volume fraction of 0.003 [19]. The L-shaped sections (CDE and FGH) are maintained at lower temperature and concentration ($\theta = 0, C = 0$), while the bottom surface (AB) is subjected to constant high temperature and concentration boundary conditions ($\theta = 1, C = 1$). The remaining walls are adiabatic and impermeable to mass transfer ($\frac{\partial \theta}{\partial n} = 0$, $\frac{\partial C}{\partial n} = 0$). No-slip conditions are applied to all boundary surfaces ($U = V = 0$).

The mathematical model follows a uniform distribution of MWCNT and Fe_3O_4 nanoparticles in water, which is considered Newtonian, incompressible, and exhibits steady, laminar flow. All thermophysical properties are constant except for density,

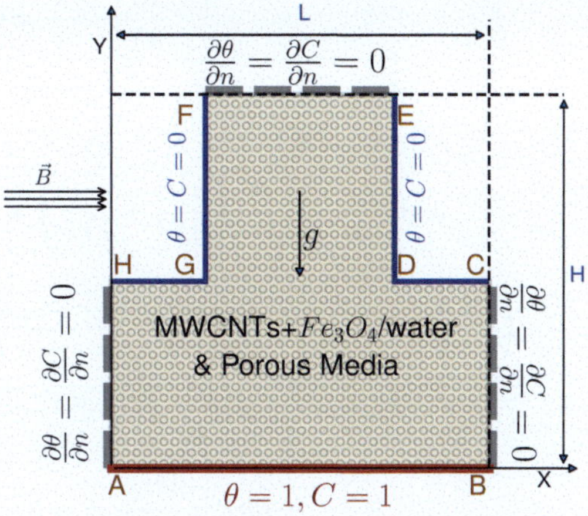

Fig. 1 Graphical representation of the physical domain with boundary conditions

which varies according to the Boussinesq approximation. The model presumes local thermal equilibrium between the fluid, matrix, and nanoparticles. Thermal radiation is considered significant only in the y-direction. Additionally, the model shows an influence of uniform magnetic field and a first-order homogeneous chemical reaction.

A 2D mathematical model is formulated based on the Darcy-Forchheimer-Brinkman equations, coupled with energy and concentration equations. This framework enables a comprehensive analysis of fluid flow, heat, and mass transfer within the system. The governing equations in the non-dimensional form are presented as follows.

The following non-dimensional parameters are employed to obtain the dimensionless governing equation [8, 20, 21]:

$$
\begin{aligned}
&X = \frac{x}{L}, Y = \frac{y}{L}, U = \frac{uL}{\alpha_f}, V = \frac{vL}{\alpha_f}, \theta = \frac{T - T_c}{T_h - T_c},\\
&P = \frac{(pL^2)}{\rho_f \alpha_L^2}, C = \frac{C^{'} - C_l^{'}}{C_h^{'} - C_l^{'}}, Pr = \frac{\nu_f}{\alpha_f}, Le = \frac{\alpha_f}{D_m},\\
&Ra = \frac{gL^3 \rho_f \beta_f (T_h - T_f)}{\mu_f \alpha_f}, Da = \frac{K}{L^2}, Du = \frac{D_m K_t \nabla C^{'}}{C_s C_p \alpha_f \nabla T},\\
&Sr = \frac{D_m K_t \nabla T}{T_m \alpha_f \nabla C^{'}}, N = \frac{\beta_{C_{nf}}}{\beta_{T_n f}} \frac{(C_h^{'} - C_l^{'})}{(T_h - T_c)}
\end{aligned}
\tag{1}
$$

The following non-dimensional form of governing equation represents the complete set of mathematical model containing continuity, momentum, energy, and solute transfer equation [8, 20, 21]:

$$\frac{\partial U}{\partial X}+\frac{\partial V}{\partial Y}=0, \tag{2}$$

$$\begin{aligned}\frac{1}{\epsilon^2}\left(U\frac{\partial U}{\partial X}+V\frac{\partial U}{\partial Y}\right)&=-\frac{\rho_f}{\rho_{nf}}\frac{\partial p}{\partial X}-\frac{\rho_f}{\rho_{nf}}\frac{\mu_{nf}}{\mu_f}\frac{Pr}{Da}U\\&+\frac{\rho_f}{\rho_{nf}}\frac{\mu_{nf}}{\mu_f}\frac{Pr}{\epsilon}\left(\frac{\partial^2 U}{\partial X^2}+\frac{\partial^2 U}{\partial Y^2}\right)-\frac{1.75}{\sqrt{150}}\frac{\sqrt{U^2+V^2}}{\sqrt{Da}}\frac{U}{\sqrt{\epsilon^3}},\end{aligned} \tag{3}$$

$$\begin{aligned}\frac{1}{\epsilon^2}\left(U\frac{\partial V}{\partial X}+V\frac{\partial V}{\partial Y}\right)=&-\frac{\rho_f}{\rho_{nf}}\frac{\partial p}{\partial Y}-\frac{\rho_f}{\rho_{nf}}\frac{\mu_{nf}}{\mu_f}\frac{Pr}{Da}V\\&+\frac{\rho_f}{\rho_{nf}}\frac{\mu_{nf}}{\mu_f}\frac{Pr}{\epsilon}\left(\frac{\partial^2 V}{\partial X^2}+\frac{\partial^2 V}{\partial Y^2}\right)-\frac{1.75}{\sqrt{150}}\frac{\sqrt{U^2+V^2}}{\sqrt{Da}}\frac{V}{\sqrt{\epsilon^3}}\\&+RaPr\frac{(\rho\beta)_{nf}}{\rho_{nf}\beta_f}(\theta+NC)-Ha^2Pr\left(\frac{\rho_f}{\rho_{nf}}\right)\left(\frac{\sigma_{nf}}{\sigma_f}\right)V,\end{aligned} \tag{4}$$

$$U\frac{\partial\theta}{\partial X}+V\frac{\partial\theta}{\partial Y}=\frac{\alpha_{nf}}{\alpha_f}\left(\frac{\partial^2\theta}{\partial X^2}+\frac{\partial^2\theta}{\partial Y^2}\right)+Du\left(\frac{\partial^2 C}{\partial X^2}+\frac{\partial^2 C}{\partial Y^2}\right)+\left(\frac{4}{3}\frac{\alpha_{nf}}{\alpha_f}\frac{k_f}{k_{nf}}\right)Rd\frac{\partial^2\theta}{\partial Y^2}, \tag{5}$$

$$U\frac{\partial C}{\partial X}+V\frac{\partial C}{\partial Y}=\frac{1}{Le}\left(\frac{\partial^2 C}{\partial X^2}+\frac{\partial^2 C}{\partial Y^2}\right)+Sr\left(\frac{\partial^2\theta}{\partial X^2}+\frac{\partial^2\theta}{\partial Y^2}\right)+C_r\cdot Pr\cdot C. \tag{6}$$

The equations use dimensionless variables like X and Y for horizontal and vertical positions, U and V for velocities in those directions, C for concentration, theta for temperature, and P for pressure. Additional parameters include Le (Lewis number), N (buoyancy ratio), Ra (Rayleigh number), Du (Dufour number), Pr (Prandtl number), Sr (Soret number), Da (Darcy number), and Cr (chemical reaction parameter).

3 Numerical Solution Methodology

The continuity equation ensures mass conservation and acts as a constraint for determining pressure distribution within the momentum equations. To handle the pressure terms arising from the momentum balance equations, a penalty FEM technique is used. This method introduces a penalty parameter ($\gamma = 10^7$) to enforce the incompressibility constraint, allowing for an approximation of the pressure term $P = -\gamma\left(\frac{\partial U}{\partial X}+\frac{\partial V}{\partial Y}\right)$. Moreover, a higher value of $\gamma = 10^7$ effectively satisfies the continuity Eq. (2) for numerical simulation process.

The momentum Eqs. (3) and (4) further become the following equation after utilizing the pressure term:

$$\frac{1}{\epsilon^2}\left(U\frac{\partial U}{\partial X}+V\frac{\partial U}{\partial Y}\right)=\gamma\frac{\rho_f}{\rho_{nf}}\frac{\partial}{\partial X}\left(\frac{\partial U}{\partial X}+\frac{\partial V}{\partial Y}\right)-\frac{\rho_f}{\rho_{nf}}\frac{\mu_{nf}}{\mu_f}\frac{Pr}{Da}U$$
$$+\frac{\rho_f}{\rho_{nf}}\frac{\mu_{nf}}{\mu_f}\frac{Pr}{\epsilon}\left(\frac{\partial^2 U}{\partial X^2}+\frac{\partial^2 U}{\partial Y^2}\right)$$
$$-\frac{1.75}{\sqrt{150}}\frac{\sqrt{U^2+V^2}}{\sqrt{Da}}\frac{U}{\sqrt{\epsilon^3}} \tag{7}$$

$$\frac{1}{\epsilon^2}\left(U\frac{\partial V}{\partial X}+V\frac{\partial V}{\partial Y}\right)=\gamma\frac{\rho_f}{\rho_{nf}}\frac{\partial}{\partial Y}\left(\frac{\partial U}{\partial X}+\frac{\partial V}{\partial Y}\right)-\frac{\rho_f}{\rho_{nf}}\frac{\mu_{nf}}{\mu_f}\frac{Pr}{Da}V$$
$$+\frac{\rho_f}{\rho_{nf}}\frac{\mu_{nf}}{\mu_f}\frac{Pr}{\epsilon}\left(\frac{\partial^2 V}{\partial X^2}+\frac{\partial^2 V}{\partial Y^2}\right)$$
$$-\frac{1.75}{\sqrt{150}}\frac{\sqrt{U^2+V^2}}{\sqrt{Da}}\frac{V}{\sqrt{\epsilon^3}}+RaPr\frac{(\rho\beta)_{nf}}{\rho_{nf}\beta_f}(\theta+NC)-Ha^2Pr\left(\frac{\rho_f}{\rho_{nf}}\right)\left(\frac{\sigma_{nf}}{\sigma_f}\right)V \tag{8}$$

The system of governing equations (Eqs. (5), (6), (7), and (8)) with boundary conditions of the model is numerically simulated using the Galerkin FEM. The bilinear basis functions were used to represent the control parameters, and the Newton-Raphson iteration algorithm was employed to handle the non-linear residual terms. Convergence of numerical approach has achieved when the relative errors for each flow variable satisfied $\frac{|\tau^{i+1}-\tau^i|}{|\tau^i|} < \eta$, where iteration number is given by i, τ denotes the flow variables, and η is the error tolerance ($\eta = 10^{-6}$). Additional computational approach can be found in the research paper [22].

The total mean Nusselt number (Nu_m) and total Sherwood number (Sh_m) monitor the convective heat and mass transfer phenomena, which are expressed by the following equation:

$$Nu_m=\int_{\partial S} Nu_l \partial S, \quad Sh_m=\int_{\partial S} Sh_l \partial S$$

where $Nu_l = \frac{hL}{k_f} + Nu_{Rd} = -\left(\frac{k_{nf}}{k_f}\frac{\partial\theta}{\partial n} + \frac{4}{3}Rd\frac{\partial\theta}{\partial n}\right)$ and $Sh_l = -\frac{\partial C}{\partial n}$ refer the local Nusselt and Sherwood number, respectively.

3.1 Thermophysical Properties of Hybrid Nanofluid

The thermophysical properties of the hybrid nanofluid were obtained from experimental data at a constant nanoparticle volume fraction of 0.003 [19]. Density variation, heat capacity, thermal expansion coefficient, electrical conductivity, and thermal diffusivity were calculated using correlations reported in research papers [23, 24]:

The significant disparity in electrical conductivity between MWCNTs of $O(10^{-7})$ and Fe_3O_4 of $O(10^4)$ shows a simple average unsuitable for reflecting the hybrid nanofluid's conductivity. Consequently, the value of $\sigma(Fe_3O_4)$ is adopted for σ_{nf} [23, 28].

4 ANN Modeling in Parameter Estimation

4.1 Mathematical Modeling of ANN

The computational ability of ANN has propelled deep learning (DL) into a prominent subfield of artificial intelligence [29]. DL's influence permeates diverse sectors, including object detection, image processing, autonomous systems, speech recognition, healthcare, and finance [10, 30]. ANNs excel at uncovering intricate patterns within data, making them indispensable tools for scientific machine learning, classification, regression, and feature extraction. Among the various ANN architectures, feedforward neural networks (FNNs) process information sequentially, establishing complex nonlinear mappings between inputs and outputs [31].

This study utilizes a fully connected neural network (FNN) to model complex relationships and predict values for Nu_m, $Be_{\theta,tot}$, and S_{tot} based on input data for flow parameters (Ra, Da, ϵ, Rd, and Ha). The basic architecture of the FNN is represented in Fig. 3a, consisting of neurons arranged in three different layers (input, hidden, and output layers). In the diagram, the colored nodes illustrate the neurons of FNN, having adjacent layers initialized with connection weights. For instance, weights w^l_{jk} refer to the connections weight from node (neural) kth in hidden layer $l-1$th to node j in hidden layer l. Additionally, each node in the layers (except the input layer) has an associated bias, denoted by b^l_j for the jth neuron in hidden layer l in Fig. 3a. Notably, there are no connections between nodes within the same layer. The input nodes $(x_1, \cdots, x_n)$ process the training data multiplied with the weighted values into the hidden layer, ultimately predicting the outputs $(y_1, \cdots, y_m)$ through the output layer. This interconnected structure enables the FNN to map complex relationships from R^n to R^m in real space.

The weighted inputs during forward-propagation move forward through the hidden layer l, derived from the previous hidden layer $l-1$, is represented by following expression:

$$a^l_j = \sum_k w^l_{jk} y^{l-1}_k + b^i_j \tag{9}$$

Here, the weighted input a^i_j received by the jth node lth hidden layer, with y^{l-1}_k representing the output value of neuron k in the preceding layer $l-1$. To introduce non-linearity into the neural network, activation functions (e.g., Tanh, ReLU, Sigmoid) are applied to the output of each neuron. Denoting the activation function as σ, the output of the jth neuron in the lth hidden layer can be expressed as

$$y_j^l = \sigma(a_j^l) = \sigma\left(\sum_k w_{jk}^l y_k^{l-1} + b_j^l\right)$$

Model parameters, including weights and biases, are optimized through backpropagation coupled with an appropriate optimization algorithm. For a detailed explanation of the backpropagation process, readers are directed to relevant research papers [31, 32].

4.2 Parametric Influence Modeling and Prediction Using ANN

This study leverages the computational power of an ANN to approximate the values of Nu_m and Sh_m based on the input flow parameters: Ra, ϵ, Da, Le, Ha, Rd, N, Sr, C_r, and Du. The structure of the fully connected neural network (FNN), shown in Fig. 3a, consists of 10 neurons in the input layer and 3 neurons in the output layer, with the number of neurons in the hidden layers adjustable to achieve the desired solution accuracy. All neurons, except those in the output layer, use the hyperbolic tangent (Tanh) function as an activation function, while the output layer neurons employ a linear activation function. The model's weights and biases are optimized using the ADAM algorithm during backpropagation, with a fixed learning rate of 0.006. The objective is to minimize the mean squared error (MSE) loss function. The numerical output of MSE is obtained from the averaged square of the desired target values (T_i) and the predicted output of ANN (O_i), which is expressed as follows:

$$MSE = \frac{\sum_{i=1}^{K} (O_i - T_i)^2}{K}$$

The index K represents the individual data samples. The training process involves iterative minimization of the loss function across the entire training dataset to optimize the ANN's weights and biases.

5 Grid Independence and Model Validation Study

A grid independence study was conducted to determine the optimal grid size for the simulations. Table 1 presents the variation of Nusselt and Sherwood numbers with increasing grid refinement (G1–G7). Minimal changes in these parameters were observed for grid sizes G5 and above, leading to the selection of grid G6 for subsequent simulations.

Table 1 Variation of Nu_m and Sh_m with various grid-types $Sr = 0.5$, $Da = 10^{-3}$, $Le = 0.1$, $Ra = 10^6$, $\epsilon = 0.4$, $Rd = 1$, $Cr = 0.1$, $Ha = 25$, $Du = 0.5$.

Grid-type	Grid size	Nu_m	Sh_m
G1	2754	11.4500	2.59545
G2	3986	11.4692	2.60087
G3	6322	11.4733	2.60244
G4	8906	11.4680	2.60165
G5	9948	11.4849	2.60532
G6	11166	11.4843	2.60541
G7	17280	11.4823	2.60512

Table 2 Mean Nusselt number (Nu_m) comparison with published research papers of Cho [25], Nithiarasu et al. [26], and Singh et al. [27] at different set of flow parameters

Ra	Present study	Cho [25]	$Da = 10^{-4}$, $\epsilon = 0.4$	
			Nithiarasu et al. [26]	Singh et al. [27]
10^3	1.007	1.077	1.01	1.01
10^4	1.408	1.360	1.408	1.361
10^5	3.166	2.994	2.983	3.067
5×10^5	5.238	4.995	4.99	5.065

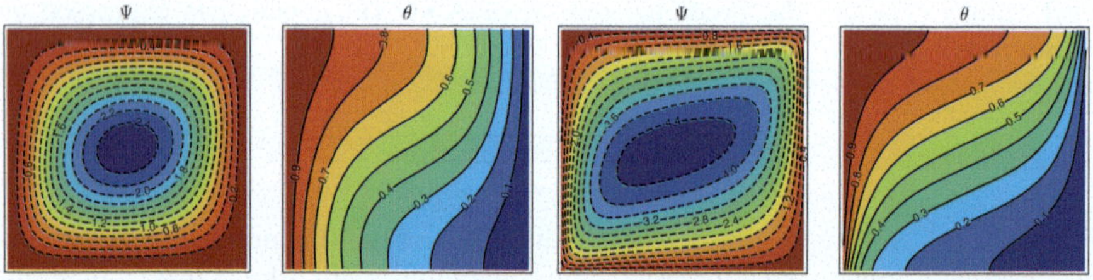

(a) $|\Psi|_{max} = 2.56$ (b) $|\Psi|_{max} = 4.66179$

Fig. 2 Streamlines (Ψ) and isotherms (θ) results validation with the results of published research papers [26] for (a) $Da = 10^{-6}$, $Ra = 10^8$, $\epsilon = 0.8$ and (b) $Da = 10^{-2}$, $Ra = 10^4$, $\epsilon = 0.6$

Model validation was achieved through comparisons with published data [25–27]. Table 2 presents a quantitative comparison of mean Nusselt number for a fixed Prandtl number (Pr = 1) with the aforementioned studies. Additionally, qualitative comparisons of streamlines and isotherms (Fig. 2) with the works of Nithiarasu et al. [26] and Basak et al. [33] were performed. Excellent agreement between the present results and published data across various flow parameters confirms the accuracy of the developed code.

6 Results and Discussion

A thorough numerical study was conducted to explore double-diffusive natural convection in a porous enclosure (inverted T-shaped). The simulation covered an extensive range of parameters, including Rayleigh number ($Ra = 10^3$ to 10^6), Darcy number ($Da = 10^{-5}$ to 10^{-2}), porosity ($\epsilon = 0.1$ to 0.9), radiation parameter ($Rd = 1$ to 5), Hartmann number ($Ha = 0$ to 100), Lewis number ($Le = 0.1$ to 0.6), buoyancy ratio ($N = -2$ to 2), chemical reaction rate ($C_r = 0.1$ to 0.9), Soret coefficient ($Sr = 0.1$ to 0.9), and Dufour coefficient ($Du = 0.1$ to 0.9).

A dataset of 155 samples, comprising both input and output parameters from numerical simulations, was utilized for training. This dataset was partitioned into training (80%), testing (10%), and validation (10%) sets. To normalize the data, Standard scaling (mean removal and unit variance scaling) was applied. Python libraries, including Scikit-learn and PyTorch, were employed for data preprocessing and model training. Figure 3b outlines the training process of the ANN-model explicitly.

The ANN model employed in this study consists of three hidden layers, each with 100 neurons. The ANN model's performance, as measured by mean squared error (MSE), converged after 400 epochs, achieving an accuracy of 98.87%. Figure 4a illustrates the MSE for training, testing, and validation datasets. Correlation analysis between predicted and experimental data, presented in Fig. 4b, reveals a strong correlation coefficient of approximately 1 for all datasets, indicating excellent agreement between the model and experimental observations. To ensure the reliability of the ANN model, its outputs have been compared with those of FEM simulations. Figures 5 and 6 present the comparison of Nu_m and Sh_m, respectively, obtained from both ANN predictions and FEM results, across various Ra values and a range of flow parameters (Rd, Da, ϵ, N, Ha, Le, Sr, C_r, & Du).

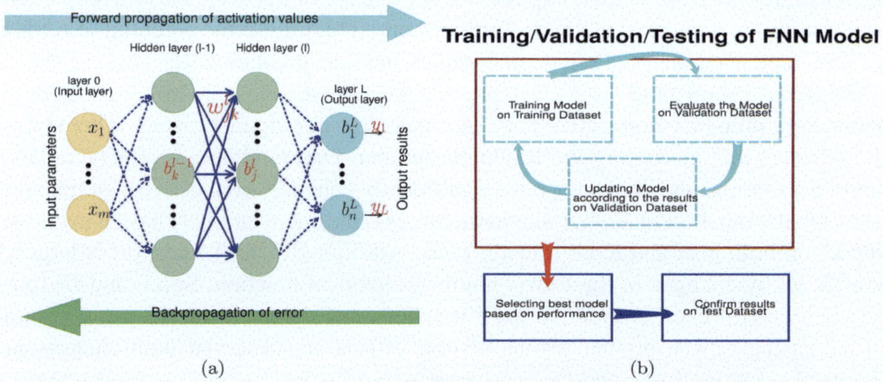

Fig. 3 Schematic representation of **a** Feedforward neural network (FNN) architecture and **b** FNN training process flow

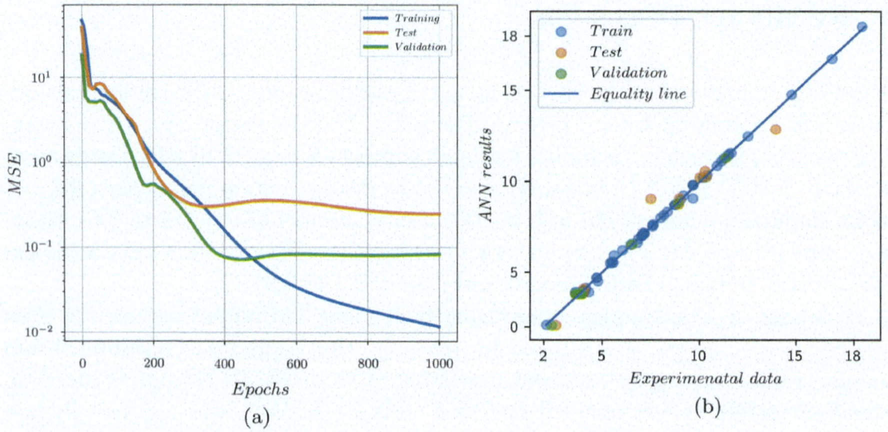

Fig. 4 Graphical representation of **a** MSE loss variation during ANN model training for training, testing, and validation datasets, and **b** comparison of ANN predicted values with experimental data

These results show a strong correlation between the predictions of ANN and FEM outcomes within the investigated parameter range, highlighting the ANN's ability to efficiently predict results with limited experimental data. Additionally, the ANN accurately approximates parametric outcomes with minimal computational expense, emphasizing the effectiveness and precision of deep learning techniques in predicting Nu_m and Sh_m, which supports their broader adoption.

FEM simulations required approximately 26 h on a standard desktop computer, while ANN training with the provided datasets was completed in only 5 min. The trained ANN accurately replicated results for a wide range of flow parameters in fractions of a second, compared to approximately 10 minutes for FEM simulations. This demonstrates the significant computational efficiency of the deep learning approach over traditional numerical methods. Furthermore, it highlights the potential of ANNs for rapid and accurate predictions in complex mathematical models.

The results shown in Figs. 5 and 6 provide valuable insights into how various flow parameters influence convective heat and mass transfer characteristics. The analysis indicates that increasing the Rayleigh number, Darcy number, porosity values, thermal radiation, and buoyancy ratio significantly enhances convective heat transfer rates. On the other hand, higher magnetic forces (Hartmann number) have a negative impact on both heat and mass transfer rates. Additionally, heat transport is largely unaffected by changes in the Lewis number, chemical reaction, Soret, and Dufour parameters. However, convective mass transfer rates show substantial growth with an increasing Lewis number, while adverse effects are observed with changes in chemical reaction, Soret, and Dufour parameters.

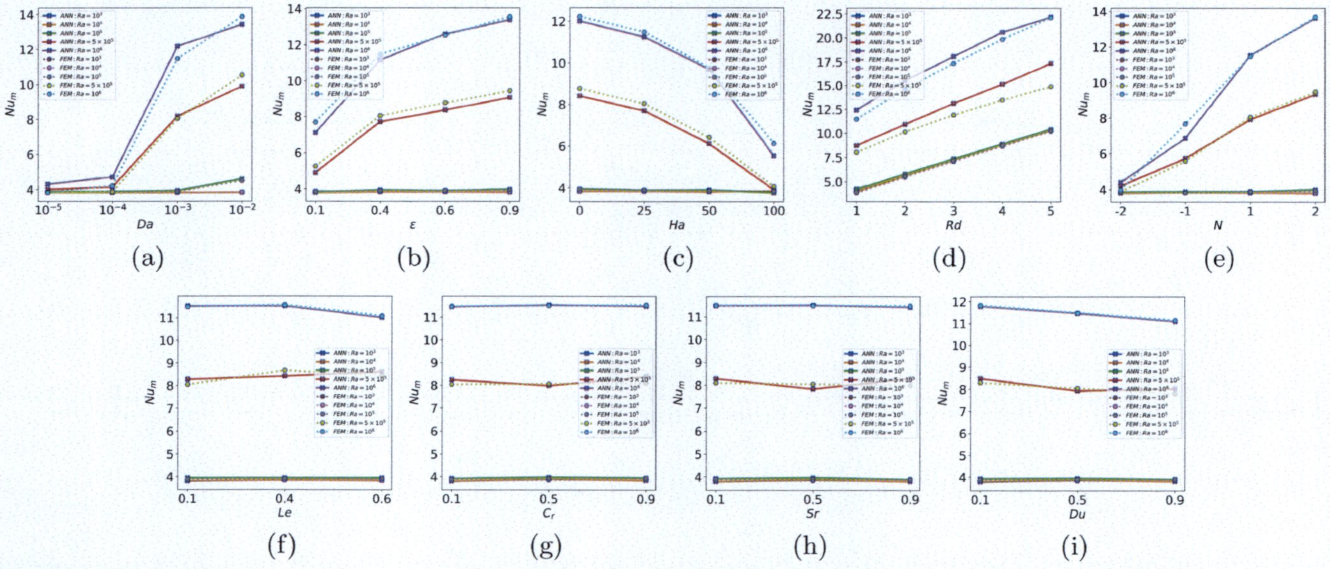

Fig. 5 Comparison of mean Nusselt number predicted by ANN and FEM models for various flow parameters: **a** Darcy number (D), **b** porosity (ϵ), **c** Hartmann number (Ha), **d** radiation parameter (Rd), **e** buoyancy ratio (N), **f** Lewis number (Le), **g** chemical reaction parameter (Cr), **h** Soret coefficient (Sr), and **i** Dufour coefficient (Du)

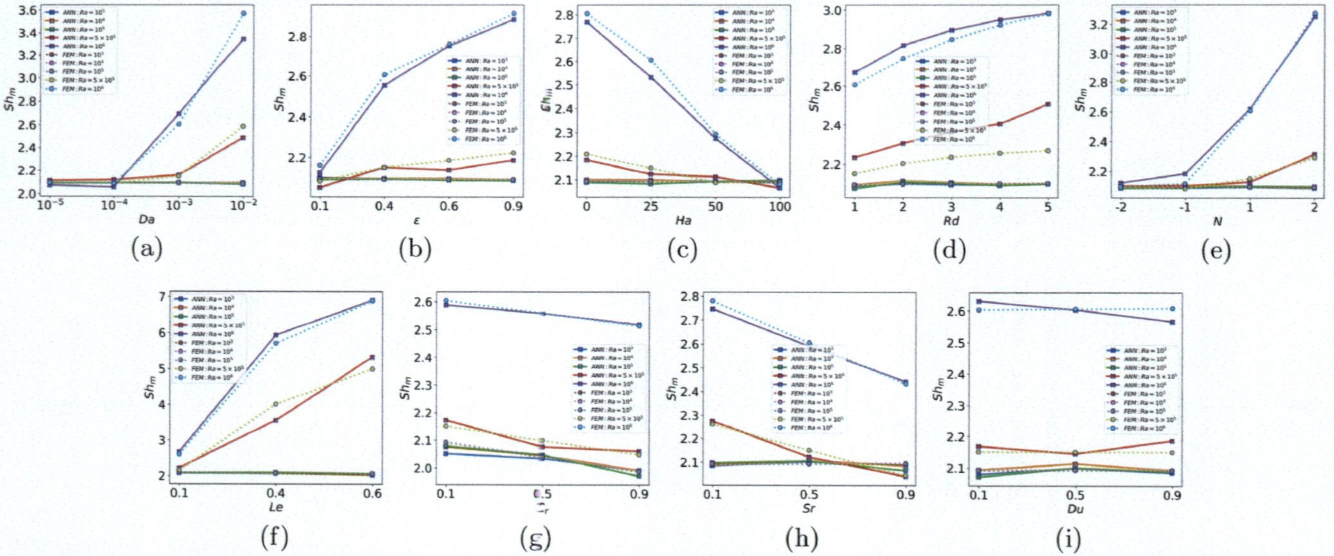

Fig. 6 Comparison of mean Sherwood number predicted by ANN and FEM models for various flow parameters: **a** Darcy number (D), **b** porosity (ϵ), **c** Hartmann number (Ha), **d** radiation parameter (Rd), **e** buoyancy ratio (N), **f** Lewis number (Le), **g** chemical reaction parameter (Cr), **h** Soret coefficient (Sr), and **i** Dufour coefficient (Du)

7 Concluding Remarks

This study employs FNN-based DL approach to explore the impact of thermal radiation, chemical reactions, magnetic fields, Soret, and Dufour effects on double-diffusive natural convection within an inverted T-shaped porous cavity. The nonlinear mathematical model coupled with ANN is employed to predict the parametric influences. Key findings are presented below.

This study validates the efficacy of an ANN-based deep learning approach in accurately predicting heat and mass transfer rates across a wide range of flow parameters. It is noticed that the trained ANN model significantly reduces computational costs while effectively capturing parametric influences with limited training data. This DL-based methodology simplifies the analysis of complex, nonlinear models governing convective heat and mass transfer. Parametric analysis reveals that increasing Rayleigh and Darcy numbers enhance heat and mass transfer through intensified convective flow and increased porous media permeability, respectively. Conversely, stronger magnetic fields suppress convective motion, leading to reduced heat and mass transfer rates. While porosity positively impacts convective transport, the influence of Lewis number, chemical reaction, Soret, and Dufour parameters on heat and mass transfer is less pronounced. These findings underscore the intricate interplay of parameters in shaping convective processes within porous media, providing valuable insights for engineering and scientific applications.

Acknowledgements The second and third authors are thankful to the SPARC Project (**Ref No. SPARC/2019-2020/P2465/SL Dt.: 28.07.2023**), funded by the Ministry of HRD, GoI, and steered by IIT-KGP in supporting and carrying out this research work. Additionally, the first author is grateful to DIAT for providing financial support and the Council of Scientific and Industrial Research (CSIR) through the International Travel Grant (**Ref. No.: TG/12386/23-HRD**) to attend FIAM-2023.

References

1. Nithiarasu, P., Seetharamu, K.N., Sundararajan, T.: Double-diffusive natural convection in an enclosure filled with fluid-saturated porous medium: a generalized non-darcy approach. Numer. Heat Transf. Part A Appl. **30**(4), 413–426 (1996)
2. Kumar, S., Rathish Kumar, B.V., Krishna Murthy, S.V.S.S.N.V.G.: Double diffusive convective flow study of a hybrid nanofluid in an inverted t-shaped porous enclosure under the influence of soret and dufour prameters. ASME J. Heat Mass Transf. **145**(10), 102501 (2023)
3. Kumar, S., Rathish Kumar, B.V., Krishna Murthy, S.V.S.S.N.V.G., Parmar, D.: Double-diffusive convective flow of hybrid nanofluid in an inverted t-shaped porous enclosure: a numerical study. Numer. Heat Transf. Part A Appl. 1–25 (2023)
4. Parmar, D., Rathish Kumar, B.V., Krishna Murthy, S.V.S.S.N.V.G., Kumar, S.: Numerical study of entropy generation in magneto-convective flow of nanofluid in porous enclosure using fractional order non-darcian model. Phys. Fluids **35**(9) (2023)
5. Al-Rashed, A.A.A.A., Oztop, H.F., Kolsi, L., Boudjemline, A., Almeshaal, M.A., Ali, M.E., Chamkha, A.: Cfd study of heat and mass transfer and entropy generation in a 3d solar distiller heated by an internal column. Int. J. Mech. Sci. **152**, 280–288 (2019)

6. Al-Mudhaf, A.F., Rashad, A.M., Ahmed, S.E., Chamkha, A.J., El-Kabeir, S.M.M.: Soret and dufour effects on unsteady double diffusive natural convection in porous trapezoidal enclosures. Int. J. Mech. Sci. **140**, 172–178 (2018)
7. Henriques, R.N., Palombo, M., Jespersen, S.N., Shemesh, N., Lundell, H., Ianuş, A.: Double diffusion encoding and applications for biomedical imaging. J. Neurosci. Methods **348**, 108989 (2021)
8. Mohamed, R.A., Abbas, I.A., Abo-Dahab, S.M.: Finite element analysis of hydromagnetic flow and heat transfer of a heat generation fluid over a surface embedded in a non-darcian porous medium in the presence of chemical reaction. Commun. Nonlinear Sci. Numer. Simul. **14**(4), 1385–1395 (2009)
9. Chiu, T.K.F., Xia, Q., Zhou, X., Chai, C.S., Cheng, M.: Systematic literature review on opportunities, challenges, and future research recommendations of artificial intelligence in education. Comput. Educ. Artif. Intell. **4**, 100118 (2023)
10. Kumar, S., Rathish Kumar, B.V., Krishna Murthy, S.V.S.S.N.V.G.: Ann-based deep collocation method for natural convection in porous media. Neural Comput. Appl. 1–17 (2024)
11. Mandal, D.K., Biswas, N., Manna, N.K., Gayen, D.K., Gorla, R.S.R., Chamkha, A.J.: Thermo-fluidic transport process in a novel m-shaped cavity packed with non-darcian porous medium and hybrid nanofluid: application of artificial neural network (ann). Phys. Fluids **34**(3) (2022)
12. Prince, H.A., Siam, M.M.H., Ghosh, A., Mamun, M.A.H.: Application of artificial intelligence on predicting the effects of buoyancy ratio on mhd double-diffusive mixed convection and entropy generation in different nanofluids and hybrid-nanofluids. J. Thermal Sci. Eng. Appl. 1–28 (2023)
13. Prince, H.A., Ghosh, A., Siam, M.M.H., Mamun, M.A.H.: Ai predicts mhd double-diffusive mixed convection and entropy generation in hybrid-nanofluids for different magnetic field inclination angles by ann. Int. J. Thermofluids 100383 (2023)
14. Filali, A., Khezzar, L., Semmari, H., Matar, O.: Application of artificial neural network for mixed convection in a square lid-driven cavity with double vertical or horizontal oriented rectangular blocks. Int. Commun. Heat Mass Transf. **129**, 105644 (2021)
15. Kumar, S., Krishna Murthy, S.V.S.S.N.V.G., Rathish Kumar, B.V., Parmar, D.: Thermo-magnetic radiative flow in porous enclosure with deep-learning parameter estimation. Int. J. Mech. Sci. **276**, 109366 (2024)
16. Seo, Y.M., Luo, K., Ha, M.Y., Park, Y.G.: Direct numerical simulation and artificial neural network modeling of heat transfer characteristics on natural convection with a sinusoidal cylinder in a long rectangular enclosure. Int. J. Heat Mass Transf. **152**, 119564 (2020)
17. Rana, P., Kumar, A., Gupta, G.: Impact of different arrangements of heated elliptical body, fins and differential heater in mhd convective transport phenomena of inclined cavity utilizing hybrid nanoliquid: Artificial neutral network prediction. Int. Commun. Heat Mass Transf. **132**, 105900 (2022)
18. Rana, P., Gupta, V., Kumar, L.: Ltne magneto-thermal stability analysis on rough surfaces utilizing hybrid nanoparticles and heat source with artificial neural network prediction. Appl. Nanosci. **13**(1), 819–838 (2023)
19. Sundar, L.S., Singh, M.K., Sousa, A.C.M.: Enhanced heat transfer and friction factor of mwcnt–fe3o4/water hybrid nanofluids. Int. Commun. Heat Mass Transf. **52**, 73–83 (2014)
20. Shoaib, M., Raja, M.A.Z., Khan, M.A.R., Farhat, I., Awan, S.E.: Neuro-computing networks for entropy generation under the influence of mhd and thermal radiation. Surf. Interfaces **25**, 101243 (2021)
21. Mohammadi, M., Gandjalikhan Nassab, S.A.: Double-diffusive convection flow with soret and dufour effects in an irregular geometry in the presence of thermal radiation. Int. Commun. Heat Mass Transf. **134**, 106026 (2022)
22. Roy, S., Basak, T.: Finite element analysis of natural convection flows in a square cavity with non-uniformly heated wall (s). Int. J. Eng. Sci. **43**(8–9), 668–680 (2005)
23. Mehryan, S.A.M., Sheremet, M.A., Soltani, M., Izadi, M.: Natural convection of magnetic hybrid nanofluid inside a double-porous medium using two-equation energy model. J. Mol. Liq. **277**, 959–970 (2019)

24. Mohebbi, R., Mehryan, S.A.M., Izadi, M., Mahian, O.: Natural convection of hybrid nanofluids inside a partitioned porous cavity for application in solar power plants. J. Thermal Anal. Calorim. **137**, 1719–1733 (2019)
25. Cho, C.-C.: Natural convection of cu-water nanofluid in enclosed cavity with porous effect and wavy surface based on energy-flux-vector visualization method. Phys. Fluids **32**(10), 103607 (2020)
26. Nithiarasu, P., Seetharamu, K.N., Sundararajan, T.: Natural convective heat transfer in a fluid saturated variable porosity medium. Int. J. Heat Mass Transf. **40**(16), 3955–3967 (1997)
27. Singh, A.K., Basak, T., Nag, A., Roy, S.: Heatlines and thermal management analysis for natural convection within inclined porous square cavities. Int. J. Heat Mass Transf. **87**, 583–597 (2015)
28. Al-Kouz, W., Aissa, A., Koulali, A., Jamshed, W., Moria, H., Nisar, K.S., Mourad, A., Abdel-Aty, A.-H., Khashan, M.M., Yahia, I.S.: Mhd darcy-forchheimer nanofluid flow and entropy optimization in an odd-shaped enclosure filled with a (mwcnt-fe3o4/water) using galerkin finite element analysis. Sci. Rep. **11**(1), 1–15 (2021)
29. LeCun, Y., Bengio, Y., Hinton, G.: Deep learning. Nature **521**(7553), 436–444 (2015)
30. Ker, J., Wang, L., Rao, J., Lim, T.: Deep learning applications in medical image analysis. IEEE Access **6**, 9375–9389 (2017)
31. Lin, J., Zhou, S., Guo, H.: A deep collocation method for heat transfer in porous media: Verification from the finite element method. J. Energy Storage **28**, 101280 (2020)
32. Lillicrap, T.P., Santoro, A., Marris, L., Akerman, C.J., Hinton, G.: Backpropagation and the brain. Nat. Rev. Neurosci. **21**(6), 335–346 (2020)
33. Basak, T., Kaluri, R.S., Balakrishnan, A.R.: Entropy generation during natural convection in a porous cavity: effect of thermal boundary conditions. Numer. Heat Transf. Part A Appl. **62**(4), 336–364 (2012)

On a $M^{[X]}/G/1$ Queue with Two Types of Random Failures, Delay in Starting the Major Repairs and Reneging During the Down Time

Kailash C. Madan

Abstract We study a single server queue with general service time distribution and two types of random system failures categorized as major or minor failures. We assume that customers arrive at the system in batches of variable size in accordance with a compound Poisson process and they are provided one by one service on a first-come, first-served basis. As soon as there is a major failure, the repairs do not start immediately resulting in delay. The delay time in starting repairs of this type of failures follows a general distribution. However, if the system fails due to a minor failure, it instantly enters a repair process with a deterministic repair time. We further assume that customers may become impatient during the breakdown periods of the system and may renege from the system. We obtain steady-state results in terms of the probability generating functions of the number of customers in the queue. Some special cases of interest are discussed, and some known results have been derived.

Keywords Batch arrivals · Random breakdowns · Repair times · Reneging · Steady state

1 Introduction

Numerous authors have studied queueing systems with service interruptions due to random failures or server vacations. We mention Levy and Yechialli [12], Takine and Sengupta [17], Madan [13], Madan et al. [14], Maraghi et al. [15], Fadhil et al. [8], among many others. In this paper, we assume that the system is subject to two types of failures categorized as major and minor failures. We assume that there is a delay in starting major repairs but no delay in starting the minor repairs. It is further assumed that during the down time of the system, customers may become impatient and may renege from the system. In the literature on queues, one finds many papers

K. C. Madan (✉)
Ahlia University, Manama, Bahrain
e-mail: kmadan@ahlia.edu.bh

R. K. Sharma et al. (eds.), *Mathematical Theory and Simulation of Scientific Problems*, Springer Proceedings in Mathematics & Statistics 487,
https://doi.org/10.1007/978-981-96-2579-6_11

on impatient customers. Customers may arrive at the system but may leave without joining the system (balking) or customers may join the queue for some time but may leave the system without getting service (reneging). For a few earlier papers on queues dealing with reneging, the reader is referred to Barrer [3], Haight [9], Ancker and Gafarian [2], and De Kok and Tijms [7]. One can find more recent papers on queues with reneging by Choi et al. [6], Zhang et al. [18], Altman and Yechiali [1], Iravani and Balcioglu [10], Boxma et al. [4], Kursk et al. [11], Choudhury and Medhi [5], and Tadj and Choudhury [16].

It may be mentioned that most of the papers on breakdown, the authors assume that the system may fail only while the system is working. However, following [14], we assume time-homogeneous random failures which means that the system may fail not only while providing service but it may also fail when it is in the idle state.

2 The Mathematical Model

The following assumptions describe the mathematical model:

- Customers arrive at the system in batches of variable size in a compound Poisson process. Let $\lambda c_i dt$ $(i = 1, 2, 3, \ldots)$ be the first-order probability that a batch of i customers arrives at the system during a short interval of time $(t, t + dt]$, where $0 \le c_i \le 1$ and $\sum_{i=1}^{\infty} c_i = 1$, $\lambda > 0$ is the mean arrival rate of batches.
- There is a single server that provides one-by-one service to customers on a "first come, first served" basis and the service time follows a general (arbitrary) distribution with distribution function $G(s)$ and density function $g(s)$. Let $\mu(x)\mathrm{d}x$ be the conditional probability density of service completion during the interval $(x, x + \mathrm{d}x]$, given that the elapsed time is x, so that

$$\mu(x) = \frac{g(x)}{1 - G(x)} \tag{1}$$

Consequently,

$$g(s) = \mu(s)e^{-\int_0^s \mu(x)dx} \tag{2}$$

- The system experiences two types of random time-homogeneous failures, i.e., it may fail at random any time, even when it is in idle state, and we categorize the two types of failures as major and minor failures. We assume that $\alpha_1 dt > 0$ and $\alpha_2 dt > 0$ are the respective probabilities that a major or a minor failure will occur during the short interval of time $(t, dt]$. Further, we assume that once the system fails, the customer whose service is interrupted, immediately comes back to the head of the queue.

- As soon as the system fails due to a major failure, the repair process does not start immediately. We assume that there is delay in starting the major repairs. We assume that the delay time "D" in starting the major repairs follows a general distribution. Let D(x) and d(x), respectively, be the distribution function and the density function of the delay time "D". Let $\delta(x)dx$ be the conditional probability of completion of delay time during the short interval of time $(t, t + dt]$, given that the elapsed time is x, therefore,

$$\delta(x) = \frac{d(x)}{1 - D(x)} \tag{3}$$

Consequently,

$$\boldsymbol{d(s) = \delta(s)e^{-\int_0^s \delta(x)dx}} \tag{4}$$

- The repair time of a major breakdown follows a general (arbitrary) distribution with distribution function Φ(r) and density function φ(r). Let β(x)dx be the conditional probability of a repair completion during the interval (x, x + dx] given that the elapsed repair time is x, so that

$$\beta(x) = \frac{\phi(x)}{1 - \Phi(x)} \tag{5}$$

Consequently,

$$\phi(r) = \beta(r)e^{-\int_0^r \beta(x)dx} \tag{6}$$

- The repair time of a minor breakdown is assumed to be deterministic with constant mean repair time $d > 0$.
- We further assume that due to the service interruptions caused by breakdowns, customers may become impatient during breakdown periods and may renege from the system. Reneging is assumed to occur according to a Poisson stream with mean reneging rate $\eta > 0$.
- Various stochastic processes involved in the system are independent of each other.

3 Definitions and Notations

We define.

$P_n(x, t)$: Probability that at time t, the server is active providing service and there are n ($n \geq 0$) customers in the queue excluding the one being served and the elapsed service time of this customer is x. Consequently, $P_n(t) = \int_0^\infty P_n(x, t)dx$ denotes the probability that at time t there are n customers in the queue excluding the one in service irrespective of the value of x.

$D_n(x, t)$: Probability that at time t, there are $n \geq 0$ customers waiting in the queue and the server is in the state of a major failure and is waiting for the repairs to start with elapsed time x. Accordingly, $D_n(t) = \int_0^\infty D_n(x, t)dx$ denotes the probability that there are $n \geq 0$ customers waiting in the queue and the server is in the state of a major failure and is waiting for the repairs to start irrespective of the value of x.

$Q_n^{(1)}(x, t)$: Probability that at time t, there are $n \geq 0$ customers in the queue and the server is under repairs of a major failure with elapsed repair time x. Consequently, $Q_n^{(1)}(t) = \int_0^\infty Q_n^{(1)}(x, t)dx$ denotes the probability that at time t there are n customers in the queue and the system is under repair irrespective of the value of x.

$Q_n^{(2)}(t)$: Probability that at time t, the system is under deterministic repairs of a minor failure and there are $n(n \geq 0)$ customers in the queue waiting for service.

Y(t): Probability that at time t, there are no customers in the system and the server is idle but available in the system.

4 Steady-State Equations Governing the System

Assuming that the steady state exists, we denote the steady-state probabilities corresponding to the probabilities defined above as follows:

$$\lim_{t\to\infty} P_n(x, t) = P_n(x) \lim_{t\to\infty} P_n(t) = \lim_{t\to\infty} \int_0^\infty P_n(x, t)dx = P_n$$

$$\lim_{t\to\infty} D_n(x, t) = D_n(x), \lim_{t\to\infty} D_n(t) = \lim_{t\to\infty} \int_0^\infty D_n(x, t)dx = D_n$$

$$\lim_{t\to\infty} Q_n^{(1)}(x, t) = Q_n^{(1)}{}_n(x), \lim_{t\to\infty} Q_n^{(1)}(t) = \lim_{t\to\infty} \int_0^\infty Q_n^{(1)}(x, t)dx = Q_n^{(1)}$$

$$\lim_{t\to\infty} Q_n^{(2)}(t) = Q_n^{(2)} and \lim_{t\to\infty} Y(t) = Y \tag{7}$$

Further, we assume that k_r is the probability of r arrivals during the repair time "d" of a type 2 failure. Therefore,

$$k_r = \frac{exp(\lambda d)(\lambda d)^r}{r!}, r = 0, 1, 2, 3, . \tag{8}$$

Then, connecting states of the system at time $t + dt$ with those at time t and then taking limit as $t \to \infty$, we obtain the following set of steady-state equations governing the system:

$$\frac{d}{dx}P_n(x) + (\lambda + \mu(x) + \alpha_1 + \alpha_2)P_n = \lambda \sum_{i=1}^{n-1} c_i P_{n-i}(x), n \geq 1 \tag{9}$$

$$\frac{d}{dx}P_0(x) + (\lambda + \mu(x) + \alpha_1 + \alpha_2)P_0(x) = 0 \tag{10}$$

$$\frac{d}{dx}D_n(x) + (\lambda + \delta(x) +)D_n = \lambda \sum_{i=1}^{n-1} c_i D_{n-i}(x) + D_{n+1}(x), n \geq 1 \tag{11}$$

$$\frac{d}{dx}D_0(x) + (\lambda + \delta(x))D_0(x) = D_1(x) \tag{12}$$

$$\frac{d}{dx}Q_n^{(1)}((x) + (\lambda + \beta(x) + \eta)Q_n^{(1)}(x) = \lambda \sum_{i=1}^{n-1} c_i Q_{n-i}^{(1)} + \eta Q_{n+1}^{(1)}(x), n \geq 1 \tag{13}$$

$$\frac{d}{dx}Q_0^{(1)}((x) + (\lambda + \beta(x))Q_0^{(1)}(x) = \eta Q_1^{(1)}(x) \tag{14}$$

$$Q_n^{(2)} = (\alpha_2 + \eta)P_{n-1}, n \geq 1 \tag{15}$$

$$Q_0^{(2)} = \alpha_2 Y \tag{16}$$

$$(\lambda + \alpha_1 + \alpha_2)Y = \int_0^\infty P_0(x)\mu(x)dx + \int_0^\infty Q_0^{(1)}(x)\beta(x)dx \tag{17}$$

The above equations are to be solved subject to the following boundary conditions:

$$P_n(0) = \int_0^\infty P_{n+1}(x)\mu(x)dx + \int_0^\infty Q_{n+1}^{(1)}(x)\beta(x)dx + \lambda c_{n+1}Y,$$
$$+\left(Q_n^{(2)}k_1 + Q_{n-1}^{(2)}k_2 + Q_{n-2}^{(2)}k_3 + \ldots + Q_1^{(2)}k_n\right), n \geq 1 \tag{18}$$

$$P_0(0) = \int_0^\infty P_1(x)\mu(x)dx + \int_0^\infty Q_1^{(1)}(x)\beta(x)dx + \lambda c_1 Y,$$
$$+\left(Q_0^{(2)}k_1 + Q_1^{(2)}k_0\right) \tag{19}$$

$$D_n(0) = \alpha_1 \int_0^\infty P_{n-1}(x)dx = \alpha_1 P_{n-1}, n \geq 1 \tag{20}$$

$$D_0(0) = \alpha_1 Y \tag{21}$$

$$Q_n^{(1)}(0) = \int_0^\infty D_{n-1}(x)\delta(x)dxn \geq 0 \tag{22}$$

5 Steady-State Queue Size Distribution at a Random Epoch

We define the following steady-state probability generating functions (PGFs):

$$P(x, z) = \sum\nolimits_{n=0}^{\infty} z^n P_n(x);\ P(z) = \sum\nolimits_{n=0}^{\infty} z^n P_n \tag{23}$$

$$D(x, z) = \sum\nolimits_{n=0}^{\infty} z^n D(x);\ D(z) = \sum\nolimits_{n=0}^{\infty} z^n D_n \tag{24}$$

$$Q^{(1)}(x, z) = \sum\nolimits_{n=0}^{\infty} z^n Q_n^{(1)}(x);\ Q^{(1)}(z) = \sum\nolimits_{n=0}^{\infty} z^n Q_n^{(1)} \tag{25}$$

$$Q^{(2)}(z) = \sum\nolimits_{n=0}^{\infty} z^n Q_n^{(2)} \tag{26}$$

$$C(z) = \sum_{i=1}^{\infty} z^i c_i,\ |z| \le 1 \tag{27}$$

Multiplying Eq. (9) by zn, sum over n from 1 to ∞, adding the result to (10), and using the generating functions defined in (22), we obtain

$$\frac{d}{dx} P(x, z) + (\lambda - \lambda C(z) + \mu(x) + \alpha_1 + \alpha_2) P(x, z) = 0 \tag{28}$$

Performing similar operations, Eqs. (11) and (12), on using (24) yield

$$\frac{d}{dx} D(x, z) + \left(\lambda - \lambda C(z) + \delta(x) + -\frac{}{z}\right) D(x, z) = 0 \tag{29}$$

Next, we use similar operations on Eqs. (13) and (14), and use (25). Thus, we get

$$\frac{d}{dx} Q^{(1)}(x, z) + \left(\lambda - \lambda C(z) + \beta(x) + -\frac{}{z}\right) Q^{(1)}(x, z) = 0 \tag{30}$$

Similarly, Eqs. (15) and (16) yield

$$Q^{(2)}(z) = (\alpha_2 +) z P(z) \tag{31}$$

Now, we multiply Eq. (18) by zn + 1, sum over n from1 to ∞, add the result to (19), and using the generating functions defined in (23), we obtain

$$zP(0, z) = \int_0^{\infty} P(x, z)\mu(x) dx + \int_0^{\infty} Q^{(1)}(x, z)\beta(x) dx + Q^{(2)}(z) K(z)$$

$$+ [\lambda C(z) - (\lambda + \alpha_1 + \alpha_2)] Y \tag{32}$$

Next, from (20) and (21), we get

$$D(0,z) = \alpha_1[Y + P(z)] \tag{33}$$

Finally, (22) yields

$$Q^{(1)}(0,z) = \int_0^\infty D(x,z)\delta(x)dx \tag{34}$$

We integrate (28), (29), and (30) between 0 and x and get

$$P(x,z) = P(0,z)e^{-(\lambda-\lambda C(z)+\mu(x)+\alpha_1+\alpha_2)x-\int_0^x \mu(t)dt} \tag{35}$$

$$D(x,z) = D(0,z)e^{-\left(\lambda-\lambda C(z)+\delta(x)+\eta-\frac{\eta}{z}\right)x-\int_0^x \delta(t)dt} \tag{36}$$

$$Q^{(1)}(x,z) = Q^{(1)}(0,z)e^{-\left(\lambda-\lambda C(z)+\beta(x)+\eta-\frac{\eta}{z}\right)x-\int_0^x \beta(t)dt} \tag{37}$$

where $P(0,z)$, D (0, z), and $Q^{(1)}(0,z)$ are given by (32), (33), and (34), respectively.

We again integrate Eqs. (35), (36), and (37) with respect to x by parts. Thus, we obtain

$$P(z) = P(0,z)\left[\frac{1-\overline{G}[\lambda-\lambda C(z)+\alpha_1+\alpha_2]}{(\lambda-\lambda C(z)+\alpha_1+\alpha_2)}\right] \tag{38}$$

$$D(z) = D(0,z)\left[\frac{1-\overline{D}\left[\lambda-\lambda C(z)+\eta-\frac{\eta}{z}\right]}{\left(\lambda-\lambda C(z)+\eta-\frac{\eta}{z}\right)}\right] \tag{39}$$

$$Q^{(1)}(z) = Q^{(1)}(0,z)\left[\frac{1-\overline{\varphi}\left[\lambda-\lambda C(z)+\eta-\frac{\eta}{z}\right]}{\left(\lambda-\lambda C(z)+\eta-\frac{\eta}{z}\right)}\right] \tag{40}$$

where $\overline{G}[\lambda-\lambda C(z)+\xi] = \int\limits_0^\infty e^{-(\lambda-\lambda C(z)+\xi)x}\cdot dG(x)$ is the Laplace–Stieltjes transform of the service time G(x), $\overline{D}\left[\lambda-\lambda C(z)+\eta-\frac{\eta}{z}\right] = \int_0^\infty e^{-\left(\lambda-\lambda C(z)++\eta-\frac{\eta}{z}\right)x}\cdot dD(x)$ is the Laplace–Stieltjes transform of the delay time D(x) and $\overline{\Phi}\left[\lambda-\lambda C(z)+\eta-\frac{\eta}{z}\right] = \int\limits_0^\infty e^{-\left(\lambda-\lambda C(z)+\eta-\frac{\eta}{z}\right)x}\cdot d\Phi(x)$ is the Laplace–Stieltjes transform of the repair time of a major failure $\Phi(x)$.

Next, multiplying both sides of Eqs. (35), (36), and (37) by $\mu(x)$, $\delta(x)$, and $\beta(x)$, respectively, and integrating over x we get

$$\int\limits_0^\infty P(x,z)\mu(x)dx = P(0,z)\overline{G}[\lambda-\lambda C(z)+\alpha_1+\alpha_2] \tag{41}$$

$$\int\limits_0^\infty D(x,z)\delta(x)dx = D(0,z)\,\overline{\overline{D}}\left[\lambda-\lambda C(z)+\eta-\frac{\eta}{z}\right] \tag{42}$$

$$\int_0^\infty Q^{(1)}(x,z)\beta(x)dx = Q^{(1)}(0,z)\overline{\Phi}\left[\lambda - \lambda C(z) + \eta - \frac{\eta}{z}\right] \tag{43}$$

Now, utilizing (41) and (43) into (32) and simplifying, we get

$$\left\{z - \overline{G}[\lambda - \lambda C(z) + \alpha_1 + +\alpha_2]\right\}P(0,z) = Q^{(1)}(0,z)\overline{\Phi}\left[\lambda - \lambda C(z) + \eta - \frac{\eta}{z}\right]$$

$$+Q^{(2)}(z)K(z) + [\lambda C(z) - (\lambda + \alpha_1 + \alpha_2)]Y \tag{44}$$

Next, we use (42) in to (34) and get

$$Q^{(1)}(0,z) = D(0,z)\overline{D}\left[\lambda - \lambda C(z) + \eta - \frac{\eta}{z}\right] \tag{45}$$

Utilizing (33) into (43), we obtain

$$Q^{(1)}(0,z) = \alpha_1[Y + P(z)]\overline{\overline{D}}\left[\lambda - \lambda C(z) + \eta - \frac{\eta}{z}\right] \tag{46}$$

Using (46) into (44) and simplifying, we get

$$P(0,z) = \frac{\begin{array}{c}\alpha_1[Y + P(z)]\overline{\overline{D}}[\lambda - \lambda C(z) + \eta - \frac{\eta}{z}]\overline{\Phi}[\lambda - \lambda C(z) + \eta - \frac{\eta}{z}] \\ +Q^{(2)}(z)K(z) + [\lambda C(z) - (\lambda + \alpha_1 + \alpha_2)]Y\end{array}}{\left\{z - \overline{G}[\lambda - \lambda C(z) + \alpha_1 + +\alpha_2]\right\}} \tag{47}$$

Next, using (47) into (38), we obtain

$$P(z) = \left[\frac{\begin{array}{c}\alpha_1[Y + P(z)]\overline{\overline{D}}[\lambda - \lambda C(z) + \eta - \frac{\eta}{z}]\overline{\Phi}[\lambda - \lambda C(z) + \eta - \frac{\eta}{z}] \\ +Q^{(2)}(z)K(z) + [\lambda C(z) - (\lambda + \alpha_1 + \alpha_2)]Y\end{array}}{\left\{z - \overline{G}[\lambda - \lambda C(z) + \alpha_1 + +\alpha_2]\right\}}\right]$$

$$\left[\frac{1 - \overline{G}[\lambda - \lambda C(z) + \alpha_1 + \alpha_2]}{(\lambda - \lambda C(z) + \alpha_1 + \alpha_2)}\right] \tag{48}$$

which further simplifies to

$$P(z)=\frac{\dfrac{\left[\dfrac{\begin{array}{c}\alpha_1 Y\,\bar{\bar{D}}\left[\lambda-\lambda C(z)+\eta-\frac{\eta}{2}\right]\bar{\Phi}\left[\lambda-\lambda C(z)+\eta-\frac{\eta}{2}\right]\\ +Q^{(2)}(z)K(z)+[\lambda C(z)-(\lambda+\alpha_1+\alpha_2)]Y\end{array}}{\{z-\bar{G}[\lambda-\lambda C(z)+\alpha_1++\alpha 2]\}}\right]}{\left[\frac{1-\bar{G}[\lambda-\lambda C(z)+\alpha_1+\alpha_2]}{(\lambda-\lambda C(z)+\alpha_1+\alpha_2)}\right]}}{1-\alpha_1\dfrac{\left[\frac{\bar{\bar{D}}\left[\lambda-\lambda C(z)+\eta-\frac{\eta}{z}\right]\bar{\Phi}\left[\lambda-\lambda C(z)+\eta-\frac{\eta}{z}\right]}{\{z-\bar{G}[\lambda-\lambda C(z)+\alpha_1++\alpha_2]\}}\right]}{\left[\frac{1-\bar{G}[\lambda-\lambda C(z)+\alpha_1+\alpha_2]}{(\lambda-\lambda C(z)+\alpha_1+\alpha_2)}\right]}} \tag{49}$$

Utilizing the value of P(z) from (49) into Eq. (33), we obtain

$$D(0,z)=\alpha_1\left[Y+\frac{\dfrac{\left[\dfrac{\begin{array}{c}\alpha_1 Y\;\bar{\bar{D}}\left[\lambda-\lambda C(z)+\eta-\frac{\eta}{z}\right]\bar{\Phi}\left[\lambda-\lambda C(z)+\eta-\frac{\eta}{z}\right]\\ +Q^{(2)}(z)K(z)+[\lambda C(z)-(\lambda+\alpha_1+\alpha_2)]Y\end{array}}{\{z-\bar{G}[\lambda-\lambda C(z)+\alpha_1++\alpha_2]\}}\right]}{\left[\frac{1-\bar{G}[\lambda-\lambda C(z)+\alpha_1+\alpha_2]}{(\lambda-\lambda C(z)+\alpha_1+\alpha_2)}\right]}}{1-\alpha_1\dfrac{\left[\frac{\bar{\bar{D}}\left[\lambda-\lambda C(z)+\eta-\frac{\eta}{z}\right]\bar{\Phi}\left[\lambda-\lambda C(z)+\eta-\frac{\eta}{z}\right]}{\{z-\bar{G}[\lambda-\lambda C(z)+\alpha_1++\alpha_2]\}}\right]}{\left[\frac{1-\bar{G}[\lambda-\lambda C(z)+\alpha_1+\alpha_2]}{(\lambda-\lambda C(z)+\alpha_1+\alpha_2)}\right]}}\right] \tag{50}$$

And then (39) yields

$$D(z)=\left\{\alpha_1\left[Y+\frac{\left[\frac{\begin{array}{c}\alpha_1 Y\ \bar{\bar{D}}\left[\lambda-\lambda C(z)+\eta-\frac{\eta}{z}\right]\bar{\Phi}\left[\lambda-\lambda C(z)+\eta-\frac{\eta}{z}\right]\\+Q^{(2)}(z)K(z)+\left[\lambda C(z)-(\lambda+\alpha_1+\alpha_2)\right]Y\end{array}}{\{z-\bar{G}[\lambda-\lambda C(z)+\alpha_1++\alpha_2]\}}\right]\left[\frac{1-\bar{G}[\lambda-\lambda C(z)+\alpha_1+\alpha_2]}{(\lambda-\lambda C(z)+\alpha_1+\alpha_2)}\right]}{1-\ \alpha_1\left[\frac{\bar{\bar{D}}\left[\lambda-\lambda C(z)+\eta-\frac{\eta}{z}\right]\bar{\Phi}\left[\lambda-\lambda C(z)+\eta-\frac{\eta}{z}\right]}{\{z-\bar{G}[\lambda-\lambda C(z)+\alpha_1++\alpha_2]\}}\right]\left[\frac{1-\bar{G}[\lambda-\lambda C(z)+\alpha_1+\alpha_2]}{(\lambda-\lambda C(z)+\alpha_1+\alpha_2)}\right]}\right\}\left[\frac{1-\bar{D}\left[\lambda-\lambda C(z)+\eta-\frac{\eta}{z}\right]}{\left(\lambda-\lambda C(z)+\eta-\frac{\eta}{z}\right)}\right] \tag{51}$$

Next, we use (50) in (45) and obtain

$$Q^{(1)}(0,z)=\alpha_1\left\{Y+\frac{\left[\frac{\begin{array}{c}\alpha_1 Y\bar{\bar{D}}\ \left[\lambda-\lambda C(z)+\eta-\frac{\eta}{z}\right]\bar{\Phi}\left[\lambda-\lambda C(z)+\eta-\frac{\eta}{z}\right]\\+Q^{(2)}(z)K(z)+\left[\lambda C(z)-(\lambda+\alpha_1+\alpha_2)\right]Y\end{array}}{\{z-\bar{G}[\lambda-\lambda C(z)+\alpha_1++\alpha_2]\}}\right]\left[\frac{1-\bar{G}[\lambda-\lambda C(z)+\alpha_1+\alpha_2]}{(\lambda-\lambda C(z)+\alpha_1+\alpha_2)}\right]}{1-\ \alpha_1\left[\frac{\bar{\bar{D}}\left[\lambda-\lambda C(z)+\eta-\frac{\eta}{z}\right]\bar{\Phi}\left[\lambda-\lambda C(z)+\eta-\frac{\eta}{z}\right]}{\{z-\bar{G}[\lambda-\lambda C(z)+\alpha_1++\alpha_2]\}}\right]\left[\frac{1-\bar{G}[\lambda-\lambda C(z)+\alpha_1+\alpha_2]}{(\lambda-\lambda C(z)+\alpha_1+\alpha_2)}\right]}\right\}\bar{\bar{D}}\left[\lambda-\lambda C(z)+\eta-\frac{\eta}{z}\right] \tag{52}$$

Then (40) gives

$$Q^{(1)}(z) = \alpha_1 \left\{ Y + \frac{\left[\frac{\alpha_1 Y \bar{D}\left[\lambda - \lambda C(z) + \eta - \frac{\eta}{z}\right]\bar{\Phi}\left[\lambda - \lambda C(z) + \eta - \frac{\eta}{z}\right] + Q^{(2)}(z)K(z) + [\lambda C(z) - (\lambda + \alpha_1 + \alpha_2)]\, Y}{\{z - \bar{G}[\lambda - \lambda C(z) + \alpha_1 + + \alpha_2]\}}\right] \left[\frac{1 - \bar{G}[\lambda - \lambda C(z) + \alpha_1 + \alpha_2]}{(\lambda - \lambda C(z) + \alpha_1 + \alpha_2)}\right]}{1 - \alpha_1 \left[\frac{\bar{D}\left[\lambda - \lambda C(z) + \eta - \frac{\eta}{z}\right]\bar{\Phi}\left[\lambda - \lambda C(z) + \eta - \frac{\eta}{z}\right]}{\{z - \bar{G}[\lambda - \lambda C(z) + \alpha_1 + + \alpha_2]\}}\right]\left[\frac{1 - \bar{G}[\lambda - \lambda C(z) + \alpha_1 + \alpha_2]}{(\lambda - \lambda C(z) + \alpha_1 + \alpha_2)}\right]} \right\}$$

$$\bar{D}\left[\lambda - \lambda C(z) + \eta - \frac{\eta}{z}\right]\left[\frac{1 - \bar{\varphi}\left[\lambda - \lambda C(z) + \eta - \frac{\eta}{z}\right]}{\left(\lambda - \lambda C(z) + \eta - \frac{\eta}{z}\right)}\right] \tag{53}$$

Finally, from (31), we obtain

$$Q^{(2)}(z) = (\alpha_2 + n)z \left\{ \frac{\left[\frac{\alpha_1 Y \bar{D}\left[\lambda - \lambda C(z) + \eta - \frac{\eta}{z}\right]\bar{\Phi}\left[\lambda - \lambda C(z) + \eta - \frac{\eta}{z}\right] + Q^{(2)}(z)K(z) + [\lambda C(z) - (\lambda + \alpha_1 + \alpha_2)]\, Y}{\{z - \bar{G}[\lambda - \lambda C(z) + \alpha_1 + + \alpha_2]\}}\right] \left[\frac{1 - \bar{G}[\lambda - \lambda C(z) + \alpha_1 + \alpha_2]}{(\lambda - \lambda C(z) + \alpha_1 + \alpha_2)}\right]}{1 - \alpha_1 \left[\frac{\bar{D}\left[\lambda - \lambda C(z) + \eta - \frac{\eta}{z}\right]\bar{\Phi}\left[\lambda - \lambda C(z) + \eta - \frac{\eta}{z}\right]}{\{z - \bar{G}[\lambda - \lambda C(z) + \alpha_1 + + \alpha_2]\}}\right]\left[\frac{1 - \bar{G}[\lambda - \lambda C(z) + \alpha_1 + \alpha_2]}{(\lambda - \lambda C(z) + \alpha_1 + \alpha_2)}\right]} \right\} \tag{54}$$

Thus, the above results in (49), (51), (53), and (54) we have determined all the PGFs, P(z), D(z), $Q^{(1)}(z)$, and $Q^{(2)}(z)$, respectively, in terms of the only unknown constant Y. Note that Y can be determined by the following normalizing condition:

$$P(1) + D(1) + Q^{(1)}(1) + Q^{(2)}(1) + Y = 1 \tag{55}$$

6 Some Special Cases

6.1 Case 1: Two Types of Failures with No Delay in Starting Major Repairs and Reneging During Downtime

In this case, putting $D_n(x, t) = 0$, $D(x, z) = 0$, and $D(z) = 0$ in the main results will give the results corresponding to this case.

6.2 Case 2: Only Major Failures with Delay in Starting Repairs and Reneging During Downtime

In this case, putting $\alpha_2 = 0$, $Q_n^{(2)} = 0$, and $Q^{(2)}(z) = 0$ in the main results will give the results corresponding to this case.

6.3 Case 3: Only Minor Failures with Deterministic Repair Times and Reneging During Downtime

In this case, putting $\alpha_1 = 0$, $Q_n^{(1)}(x, t) = 0$, $Q^{(1)}(x, z) = 0$, and $Q^{(1)}(z) = 0$ $D_n(x, t) = 0$, $D(x, z) = 0$, and $D(z) = 0$ in the main results will give the results corresponding to this case.

6.4 Case 4: Two Types of Failures with Delay in Starting Major Repairs and No Reneging During Downtime

In this case, putting $\eta = 0$ in the main results will give the results corresponding to this case.

6.5 Case 5: No Failures, No Delay, and No Reneging

In this case, we put $\alpha_1 = 0$, $\alpha_2 = 0$, and $\eta = 0$ in the main results and obtain

$$D(Z) = 0,\ Q^{(1)}(z) = 0, Q^{(2)}(z) = 0$$

and

$$P(z) = \frac{[\lambda C(z) - \lambda]Y}{z - \overline{G}[\lambda - \lambda C(z)]} \tag{56}$$

Then the normalizing condition

$$P(1) + Y = 1 \tag{57}$$

yields

$$Y = 1 - \lambda E(S) \tag{58}$$

where E(S) is the mean service time.

Using this value of Y from (58), Eq. (56) gives

$$P(z) = \frac{[\lambda C(z) - \lambda]1 - \lambda E(S)}{z - \overline{G}[\lambda - \lambda C(z)]} \tag{59}$$

The results (58) and (59) are known results of the Mx/G/1 queue.

7 Conclusions

We have studied a new queueing model in which the service system is subject to two types of random failures—a major failure and a minor failure. We assume that there is delay in starting repairs for a major failure, whereas the minor failure is attended to instantly. A most significant assumption added to the system is that during the duration of a failure, customers possibly get impatient and some of them might renege without getting served. The system equations have been solved in the steady state in terms of the generating functions. The results for some interesting special cases have been derived from the main results. The paper obtains new significant results which add value to the theory of queues.

Acknowledgements The author wishes to sincerely thank Ahlia University for the encouragement and support to carry out this research and complete it successfully.

References

1. Altman, E., Yechiali, U.: Analysis of customers' impatience in queue with server vacations. Queuing Syst. **52**(4), 261–279 (2006)
2. Ancker, C.J., Jr., Gafarian, A.V.: Some queuing problems with balking and reneging. Oper. Res. **11**(1), 88–100 (1963)
3. Barrer, D.Y.: Queueing with impatient customers and ordered service. Oper. Res. **5**, 650–654 (1957)
4. Boxma, O., Perry, D., Wolfgang, S., Shelley, Z.: The busy period of an M/G/1 queue with customer impatience. J. Appl. Probab. **47**, 130–145 (2010)
5. Choudhury, A., Medhi, P.: Balking and reneging in multi-server Markovian queuing systems. Int. J. Math. Oper. Res. **3**(4), 377–394 (2011)
6. Choi, B.D., et al.: M/M/1 queue with impatient customers of higher priority. Queueing Syst. **38**, 49–66 (2001)
7. De Kok, A.G., Tijms, H.C.: A queuing system with impatient customer. J. Appl. Probab. **22**, 688–696 (1985)
8. Fadhil, R., Madan, K.C., Cormac, L.: An M(X)/G/1 queue with Bernoulli schedule general vacation times, random breakdowns, general delay times and general repair time. Appl. Math. Sci. **5**(1), 35–51 (2011)
9. Haight, F.A.: Queuing with balking. Biometrika **44**, 360–369 (1957)
10. Iravani, F., Balcioglu, B.: On priority queues with impatient customers. Queueing Systems–Theory Appl. **58**(4), 39–260 (2008)
11. Kursk, L., et al.: Heavy traffic analysis for EDF queues with reneging. Ann. Appl. Probab. **21**(2), 484–545 (2011)
12. Levy, Y., Yechialli, U.: An M/M/S queue with servers' vacations. Infor **14**, 153–163 (1976)
13. Madan, K.C.: A priority queuing system with service interruptions. Stat. Neerl. **27**(3), 115–123 (1973)
14. Madan, K.C., et al.: On two parallel servers with random breakdowns. Soochow J. Math. **29**(4), 413–423 (2003)
15. Maraghi, F., et al.: Bernoulli schedule vacation queues with batch arrivals and random system breakdowns having general repair time distribution. Int. J. Oper. Res. **7**(2), 240–256 (2010)
16. Tadj, L., Choudhury, G.: The M(X)/G/1 queue with unreliable server, delayed repairs, and Bernoulli vacation schedule under T-policy. Appl. Appl. Math. **8**(2), 46–365 (2013)

17. Takine, T., Sengupta, B.: A single server queue with service interruptions. Queueing Syst. **26**(3), 285–300 (1997)
18. Zhang, Y., Yue, D., Yue, W.: Analysis of an M/M/1/N queue with balking, reneging and server vacations. Int. Symp. Its Appl. (2005)

Triple Secure Encryption Scheme Three-Channel Image Based on Hankel Transform, Geometric Transforms and Hyper-Chaotic Maps

Vrushali P. Khaladkar and Manish Kumar

Abstract Daily usage of cyberspace and many telecommunication technologies like the Internet of Things (IoT), data transmission through clouds and wireless networks have proliferated. Substantial amounts of data are getting transmitted every day and the number is astronomically huge. The data contains textual data, images, videos and audio. Securing this data which is transmitted through channels of cyberspace is essential. The presented work discusses securing data of the type videos and images. The inspiration behind this research is to furnish a proficiently assembled three-channel image encryption scheme. The proposed work presents a dynamic encryption method for three-channel images built on key generation using the Hankel transform matrix, geometric transformation matrix and hyper-chaotic map to achieve triple security. Computer-simulated results show satisfactory output towards statistical tests. The sustainability of the work proposed in this research, against diverse attacks like differential, injection of noise attacks and cutting channel attacks, is tested and the proposed encryption scheme proves it promisingly. The robustness and effectiveness of the proposed scheme are decently compared with other existing schemes and it demonstrates the practical usability of the proposed encryption scheme.

Keywords Encryption · Decryption · Hyper-chaotic maps · Hankel transform · Fibonacci Q matrix

V. P. Khaladkar (✉) · M. Kumar
Department of Mathematics, Birla Institute of Technology and Science-Pilani, Hyderabad Campus, Hyderabad, Telangana, India
e-mail: p20200511@hyderabad.bits-pilani.ac.in

V. P. Khaladkar
Department of Mathematics, Fergusson College(Autonomous), Pune, Maharashtra, India

R. K. Sharma et al. (eds.), *Mathematical Theory and Simulation of Scientific Problems*, Springer Proceedings in Mathematics & Statistics 487,
https://doi.org/10.1007/978-981-96-2579-6_12

1 Introduction

In the world of data-driven applications and information technology, the daily exchange of various types of information is in millions of bytes. Securing this data from various attacks, from non-authenticate users, is of utmost priority. The well-known technique to secure digital images is to convert images into ambiguous form using secret keys and retrieval of the plain three-channel image using the same keys. Most of the image encryption algorithms are rooted in number theoretic algorithms like the RSA algorithm, ElGamal encryption scheme, and confusion and diffusion using chaotic maps and permutation ciphers. The proposed encryption scheme for three-channel images is based on two steps: cryptographic confusion using the Hankel transform and geometric transform along with a hyper-chaotic dynamical system of dimension 6 and cryptographic diffusion using 2×2 Fibonacci $\mathcal{Q}$-matrix. Various tests are performed on the proposed encryption scheme to test its sturdiness against various attacks. Carrying out of the proposed encryption scheme is compared with several prevailing research works using statistical tests comprising information entropy analysis, Pearson correlation coefficient, histograms, scatter plots, etc. The consequences of these tests confirm the excellent security and outstanding performance of the proposed work.

2 Literature Review

Daily routine process includes transmission and storage of a huge image dataset. Usually, users choose to secure their image data on social networks. Security of patient data like medical images is necessary under healthcare networks because attacks like data cut attacks or noise attacks may lead to wrong medical diagnoses. Geographic information imaging and military images require security algorithms with high strength to prevent information from getting leaked. Generally, proprietors of digital/high-tech images prefer not to allow others to ingress their images like company logos without their consent. To reduce unauthorized access to image data from various realms, the security of images has become an utmost priority. There are mainly three ways to secure image datasets, *viz.* data hiding, image watermarking and encryption of images. Numerous researchers have proposed several algorithms to secure image data using number theoretic algorithms like RSA, ElGamal encryption and discrete logarithm problem while plenty of researchers prefer to use cryptography-based cryptographic confusion and cryptographic diffusion based on dynamical systems that show chaotic nature because of their sensitivity towards initial conditions [1, 2] and transposition ciphers for encryption of image data. DNA encryption is also one of the rapidly used techniques for images using biological DNA sequencing and scrambling it. The researchers succeed in their research of encryption algorithms which stands on the use of fractals and three-dimensional Lorenz chaotic systems [3, 4]. Author [5] presents the work that uses discrete chaotic maps along with continuous chaotic maps for confusion and diffusion processes in encryption. Multimedia security I presented in the research work by the author [6] using dynamical systems with chaotic nature. The techniques based on digital watermarking, steganography and data hiding have been used by many novelists [7–9] in their research for information security. Numerous innovations are carried out in chaotic dynamical systems to obtain hybrid chaotic systems for designing encryption algorithms [1, 10–15]. The key generation based on improved Rossler's system along with multidimensional chaotic maps and encryption based on DNA sequences is showcased in the research [16]. A productive encryption scheme based on two-dimensional cellular automata is proposed in the work [17]. The author [18] proposed how chaotic neural networks work better in image encryption along with event-induced impulsive control, impulsive control and delayed event-induced delayed impulsive control for the construction of fractals. A novel encryption technique based on the generation of seed using fractional ordered chaotic systems is designed in the work [19] which resolves the issues related to multi-faceted operations involved in low-dimensional chaotic systems and simple but multidimensional systems. The work [20] shows how hiding image by image is a successful dual algorithm for protecting visually meaningful carrier image (CAI) using a secret image (SI). Six-dimensional hyper-chaotic dynamical system with Fibonacci $\mathcal{Q}$-Matrix is used for key generation and cryptographic confusion and diffusion for protecting colour image data is recommended by the researchers [21].

The proposed encryption scheme uses three levels: key generation using Hankel matrix transform and geometric transform, first-level encryption using this key and second-level encryption using six-dimensional hyper-chaotic maps with Fibonacci $\mathcal{Q}$-Matrix. The strength and effectiveness of the proposed 3-channel image encryption are examined for multiple images of varied sizes and the tabulated results in the analysis section show the impact of the proposed scheme against differential attacks, noise injection attacks, cutting channel attacks, etc.

3 Framework for Proposed Three-Channel Image Encryption Scheme

3.1 6-D Hyper-Chaotic System

Analysis of various dynamic systems at specific seed values shows a chaotic nature and hence responses of such systems are unpredictable. The behaviour of hyper-chaotic systems of higher dimension is more complicated and unpredictable in nature than the low-dimensional systems with chaotic nature. A dynamical system with a minimum of four dimensions and a minimum of two positive Lyapunov exponents is referred to as a hyper-chaotic dynamical system. Authors [21] defined 6-D hyper-chaotic system in (1):

$$\begin{cases} \dot{z_1} = \alpha(z_2 - z_1) + z_4 - z_5 - z_6 \\ \dot{z_2} = \gamma z_1 - z_2 - z_1 z_3 \\ \dot{z_3} = -\beta z_3 + z_1 z_2 \\ \dot{z_4} = \delta z_4 - z_2 z_3 \\ \dot{z_5} = \eta z_6 + z_3 z_2 \\ \dot{z_6} = \rho z_1. \end{cases} \tag{1}$$

In Eq. (1), parameters α, β, γ, δ, η and ρ are constants while z_1, z_2, z_3, z_4, z_5 and z_6 are variables representing state of the 6-D hyper-chaotic system.

3.2 2 x 2 Fibonacci $\mathcal{Q}$-Matrix

The Fibonacci sequence $\mathscr{F}_m$, outlined as (2)

$$\mathscr{F}_m = \mathscr{F}_{m-1} + \mathscr{F}_{m-2}, \quad \text{for } m > 2, \tag{2}$$

where $\mathscr{F}_1 = \mathscr{F}_2 = 1$. Fibonacci $\mathcal{Q}$-Matrix of power m is given by

$$\mathcal{Q}^m = \begin{bmatrix} \mathscr{F}_{m+1} & \mathscr{F}_m \\ \mathscr{F}_m & \mathscr{F}_{m-1} \end{bmatrix}.$$

Here, F_m is mth Fibonacci number. Important property of the Fibonacci $\mathcal{Q}$-Matrix is given as follows:

$$Determinant(\mathcal{Q}^m) = Determinant \begin{bmatrix} \mathscr{F}_{m+1} & \mathscr{F}_m \\ \mathscr{F}_m & \mathscr{F}_{m-1} \end{bmatrix}.$$

Hence, $Determinant(\mathcal{Q}^m) = \mathscr{F}_{m+1}\mathscr{F}_{m-1} - {\mathscr{F}_m}^2 = (-1)^m$. Also, inverse of Fibonacci $\mathcal{Q}$-Matrix is obtained as follows:

$$inv(\mathcal{Q})^m = \begin{bmatrix} \mathscr{F}_{m-1} & -\mathscr{F}_m \\ -\mathscr{F}_m & \mathscr{F}_{m+1} \end{bmatrix}.$$

3.3 Hankel Matrix Transform

A finite ordered or infinite ordered square matrix is said to be a Hankel matrix if it has constants on each diagonal that are orthogonal to the main diagonal and whose entries are defined as follows (3):

$$\mathscr{H}_{p,q} = \begin{cases} 0, & p+q-1 > n \\ p+q-1, & otherwise. \end{cases} \tag{3}$$

There are several different ways to define Hankel transform, one of the special cases is a Hilbert matrix, whose (p, q)th entry is $\frac{1}{p+q-1}$, $p = 1, 2, 3, \ldots$ and $q = 1, 2, 3, \ldots$. Various special cases of Hankel matrix transform lead to different applications. In this proposed work, the Hankel transform defined is used to generate an initial key for encryption of the three-channel image.

4 Schema of Encryption and Decryption

This section discusses the proposed three-channel image encryption and decryption scheme along with key generation.

4.1 *Initial Key Generation Scheme*

In the current section, key generation for the first level of encryption using Hankel transform is proposed. The final key is generated using the Hankel transform of individual channels of the plain three-channel image along with rotation transform by 180°. This key acts as an initial key for the first level of encryption given in the flow diagram shown in Fig. 1.

5 Level-Wise Encryption

This section discusses the proposed three-channel image encryption and decryption scheme along with key generation. The first level of three-channel image encryption is implemented using XOR operation on separate channels of three-channel image with a key generated using the proposed key generation scheme. The diagrammatic representation is shown in Fig. 2.

5.1 *First Level of Encryption*

See Fig. 2.

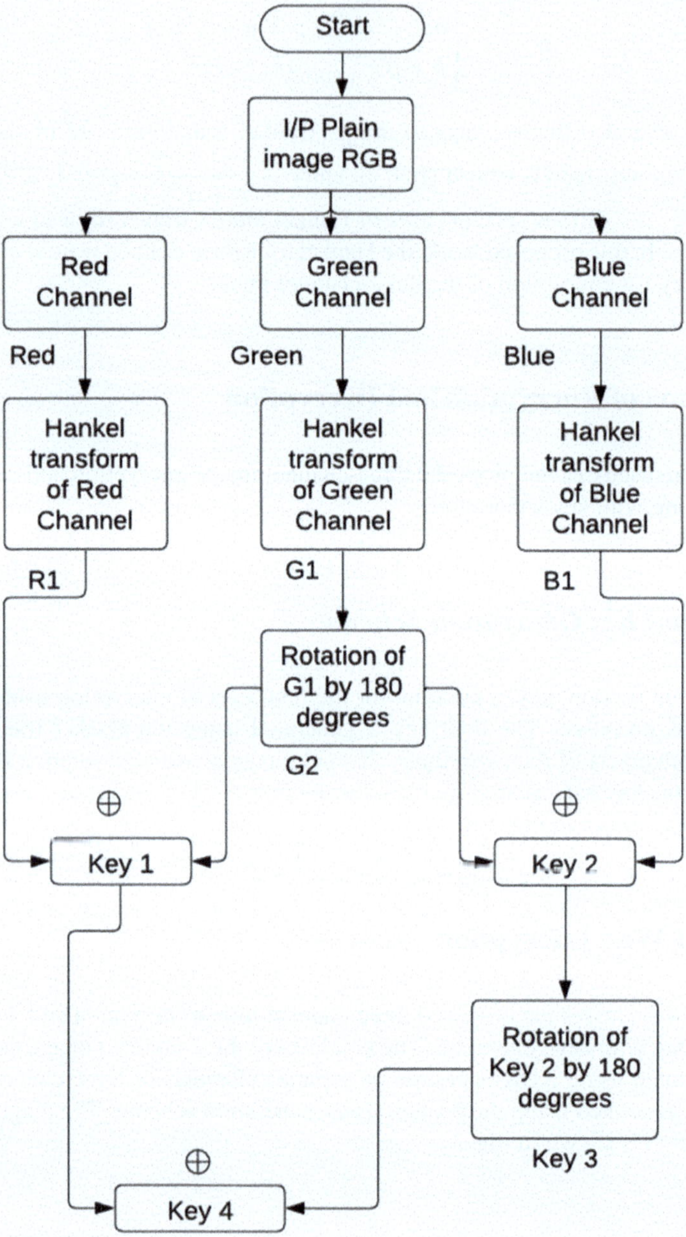

Fig. 1 Initial key generation

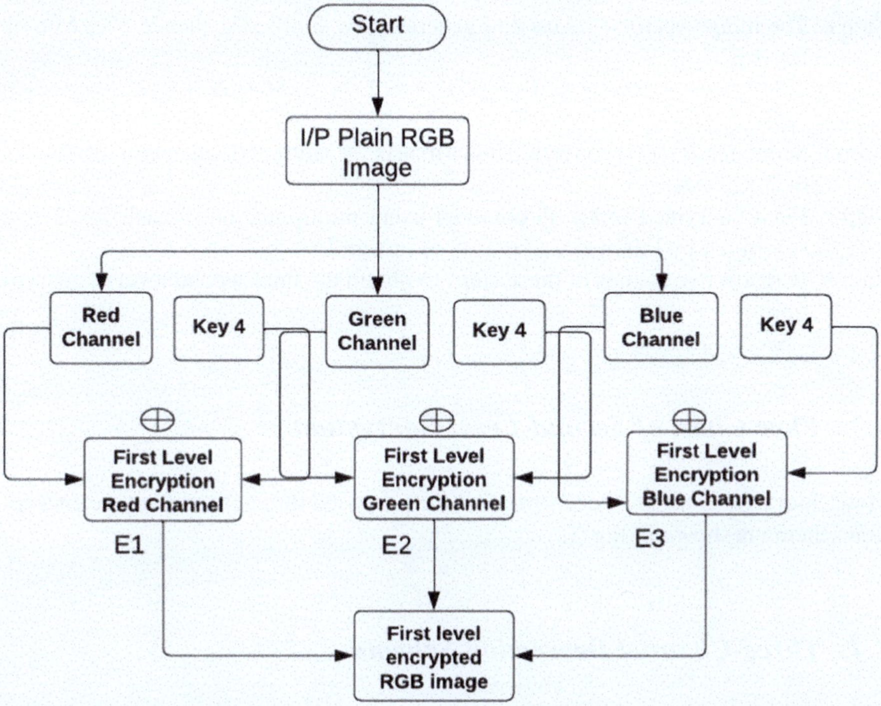

Fig. 2 First level of encryption

5.2 *Second Level of Encryption*

To achieve a second level of encryption, the following steps are used:

Step a An image array is transformed to a vector $\mathscr{V}$.

Step b Seed value of six-dimensional hyper-chaotic dynamical system is calculated as (4)

$$x_1 = \frac{\sum_{n=1}^{MN} \mathscr{V}(i) + M \times N}{2^{46} + (M \times M)}, \tag{4}$$

while x_i for $i = 2, 3, ..., 6$ are successively calculated using Eq. (5):

$$x_i = mod(x_{i-1} \times 10^6, 1). \tag{5}$$

Step c With these seed values the six-dimensional hyper-chaotic dynamical system is recursively calculated $\frac{0.9865 \times M \times N}{3}$ times to procure a sequence $\mathscr{S}_1$ of size $M \times N$.

Step d A new vector $\mathscr{S}_2$ is obtained by classifying sequence $\mathscr{S}_1$ in increasing order and returning their locations in $\mathscr{S}_2$.

Step e The image vector $\mathscr{V}$ is used to generate new shuffled sequence $\mathscr{S}_3$ given by Eq. (6):

$$\mathscr{S}_3(i) = \mathscr{V}(\mathscr{S}_2(i)), i = 1, 2, ..., M \times N. \tag{6}$$

Step f Sequence $\mathscr{S}_3(i)$ is converted into a matrix $\mathscr{M}$, and subdivided into sub-blocks of 2×2 size.

Step g Final encrypted image is obtained using multiplication of each 2×2 size sub-block by Fibonacci $\mathscr{Q}$ matrix of power 25.

Step h Perform five rounds of these steps to obtain the final second level of encryption.

5.3 Flow Chart of Second-Level Encryption

Flow chart of second-level encryption of the proposed three-channel image encryption scheme is shown in Fig. 3.

5.4 Three-Channel Decryption Scheme

Encryption using a symmetric key, public–private key and hash functions are the major three types of any data encryption scheme based on mathematical cryptography. The proposed three-channel encryption scheme is based on symmetric key generation. To retrieve the original three-channel image, the reverse encryption scheme with suitable ordering is applied. The reverse chaotic map and inverse of Fibonacci $\mathscr{Q}$ matrix are used in this reverse approach for recovering the original three-channel image. For simulation results, the Fibonacci $\mathscr{Q}$ matrix is used which is given by

$$\mathscr{Q}^{15} = \begin{bmatrix} 987 & 610 \\ 610 & 377 \end{bmatrix},$$

while for simulation results of proposed three-channel decryption scheme, inverse $(\mathscr{Q}^{15})^{-1}$ is used, where

$$(\mathscr{Q}^{15})^{-1} = \begin{pmatrix} 987 & 610 \\ 610 & 377 \end{pmatrix}.$$

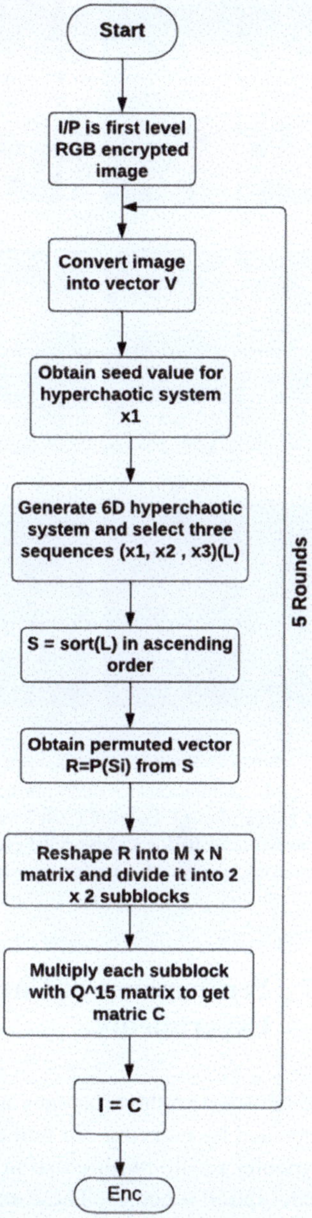

Fig. 3 Flow chart of second-level encryption of the proposed encryption scheme

6 Encryption Results

Mathematical simulation of the proposed encryption scheme is executed on MATLAB 2023a version and results are figured out in the following figure. MATLAB simulation is performed on three-channel images with various sizes and intensities (Fig. 4).

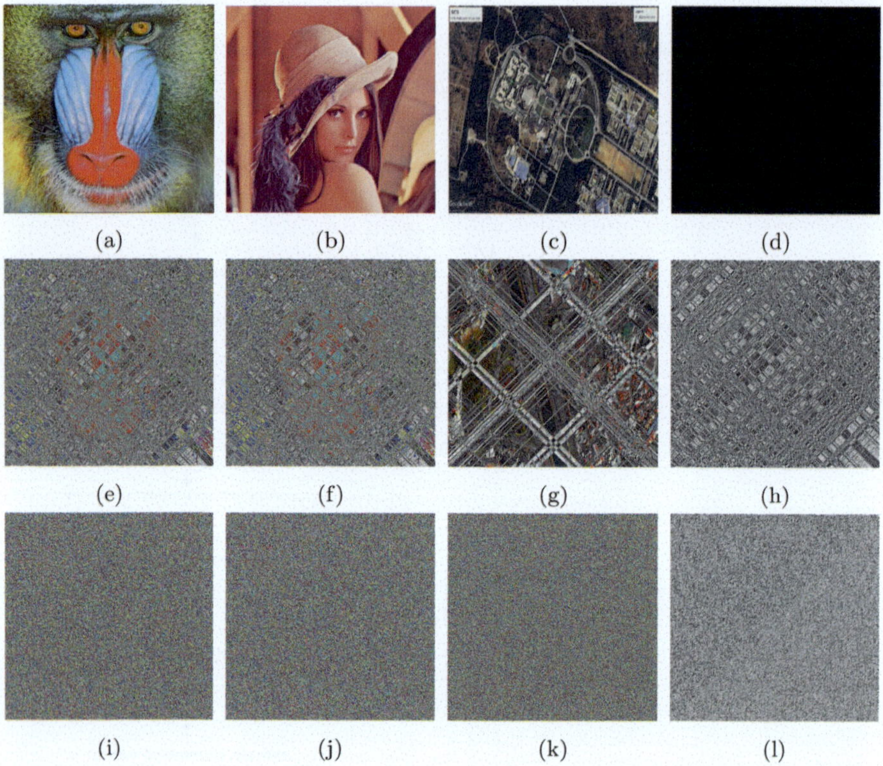

Fig. 4 Simulations for encryption and decryption: **a** Baboon (plain), **b** Lena (plain), **c** City (plain), **d** Black(plain) **e** Baboon (first-level encryption) **f** Lena (first-level encryption), **g** City (first-level encryption), **h** Black (first-level encryption), **i** Baboon (second-level encryption), **j** Lena (second-level encryption), **k** City (second-level encryption), **l** Black (second-level encryption)

7 Performance of Encryption Scheme Through Statistical Experiments

Evaluation of the efficiency and productivity of the proposed encryption scheme is verified by carrying out numerous experiments and decently collating these experimental results with existing competitive algorithms and proposed three-channel encryption schemes. These experiments involve different statistical tests like analysis through histograms, Shannon entropy test, mean square error, correlation analysis, scatter plots, etc. Also, the proposed scheme is tested for various attacks like differential attacks, trimming channel attacks, contamination attacks, blurring attacks, etc. The decryption steps of the proposed scheme show successful execution and retrieve the plain image from the cipher image.

7.1 Assessment Through Histograms

The uniform distribution of intensities after encryption proves the strength of the randomness of intensity samples of cipher space. The histogram of any 3-channel electronic image with a dynamic range from 0 to 255 is a discrete function defined as (7)

$$\text{Histogram}(r\text{th intensity value}) = \text{number of pixels comprising intensity level}(r). \quad (7)$$

Tabulated images show the histograms of 3-channel images for separated channels pre- and post-application of the proposed encryption scheme. The histogram analysis of Baboon and Lena images is shown in Fig. 5.

7.2 Assessment Through Shannon Information Entropy

The arbitrariness of any cryptographic scheme is evaluated through information entropy analysis. Information leakage is also measured using entropy analysis. Shannon entropy analysis is carried out on various 512 × 512 and 256 × 256 size images of various intensities and the results of the simulation show the highest entropy value. For any grayscale 8-bit image, the desirable value of Shannon information

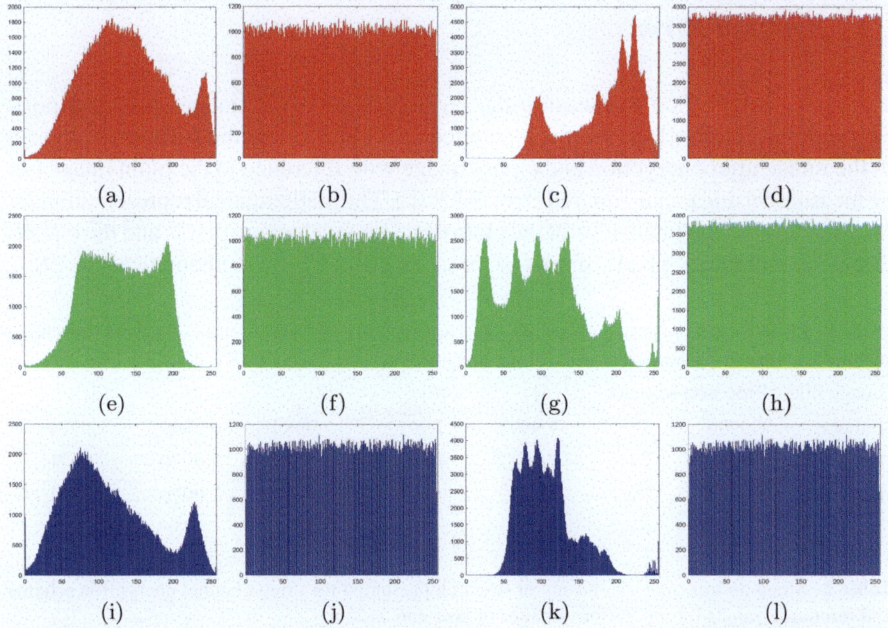

Fig. 5 Assessment through channel-wise histograms: **a** Baboon (red-plain), **b** Baboon (red-encrypted), **c** Lena (red-plain), **d** Lena (red-encrypted), **e** Baboon (green-plain), **f** Baboon (green-encrypted), **g** Lena (green-plain), **h** Lena (green-encrypted), **i** Baboon (blue-plain), **j** Baboon (blue-encrypted), **k** Lena (blue-plain), **l** Lena (blue-encrypted)

entropy is 8, the results tabulated below guarantee that the leakage of information during the encryption process is much less in the outlined three-channel image encryption scheme. Hence, the robustness of the proposed 3-channel encryption system against the entropy attack is verified agreeably. Mathematically, Shannon information entropy is given by (8)

$$I_{\text{Entropy}} = \sum_{r=0}^{2^L-1} \mathscr{P}(m_r) log_2 \frac{1}{\mathscr{P}(m_r)}, \tag{8}$$

where r = intensity values from [0, $2^L - 1$], m_r = an event such that the intensity value at specific location is r, L = number of bits used to represent the image, $\mathscr{P}(m_r)$ = probability of occurrence of m_r. For any purely random L-bit image, the probability values $\mathscr{P}(m_r)$ are almost equal and hence show Shannon's information entropy approximately to L. Post-encryption, the cipher image shows high randomness in intensities. The tabulated values (Tables 1 and 2) showcase the Shannon information entropy for the research work proposed along with decent comparison with reference values for various 3-channel images of size 512 × 512 and 256 × 256.

7.3 *Analysis of Proposed Encryption Scheme Through Mean Squared Error*

The statistical method to measure the error present in the images pre- and post-encryption is referred to as the mean-squared error (M.S.E.) method. The error present in the image after encryption and cipher image with reference to the plain image can be measured using mean-squared error (M.S.E.). The mathematical representation for M.S.E. between the intensities of the plain three-channel image pixels and the cipher-three-channel image pixels of size $\mathscr{M} \times \mathscr{N}$ is given by the mathematical Eq. (9):

Table 1 Shannon information entropy of values of research presented for three-channel encryption scheme and reference values for the various images of size 512 × 512

Image title	Proposed scheme	[4]	[6]	[7]	[22]	[8]
Baboon	7.9998	7.9972	7.9912	7.9989	7.9969	7.9974
Lena	7.9999	7.9972	7.9909	7.9972	7.9970	7.9971
Peppers	7.9995	7.9972	7.9909	7.9972	7.9970	7.9971

Table 2 Shannon information entropy of research presented for three-channel encryption scheme and reference values for the various images of size 256 × 256

Image title	Proposed scheme	[4]	[6]	[7]	[22]	[8]
Baboon	7.9992	7.9972	7.9912	7.9989	7.9969	7.9974
Lena	7.9990	7.9972	7.9909	7.9972	7.9970	7.9971
Peppers	7.9993	7.9972	7.9909	7.9972	7.9970	7.9971

$$I_{M.S.E.} = \frac{1}{\mathscr{M} \times \mathscr{N}} \sum_{y=1}^{\mathscr{N}} \sum_{x=1}^{\mathscr{M}} [\mathscr{I}_{enc}(x, y) - \mathscr{I}_{ori}(x, y)]^2, \tag{9}$$

where $\mathscr{M}$ = image row count in a plain or cipher image, $\mathscr{N}$ = image column count in a plain or cipher image, $\mathscr{I}_{enc}(x, y)$ = intensity values of the pixel at the (x, y)th location of the cipher image, $\mathscr{I}_{ori}(x, y)$ = intensity values of the pixel at (x, y)th location of the plain image. The mean-squared error for image size 512×512 between the plain three-channel image and cipher image is given in Table 3.

7.4 Assessment Through Correlation Coefficient

For any plain image, generally, the pixels located in a row, column or diagonal are strongly correlated. That is, pixels positioned in vertical(I_V), horizontal(I_H) or diagonal(I_D) direction show high correlation. The exact opposite happens for the encrypted image. In this analysis, neighbouring 10,000 pixels are chosen randomly and the Pearson correlation coefficients are computed for the vertical, horizontal and diagonal orientations. The following tabulated values (Table 4) show the correla-

Table 3 $I_{M.S.E.}$ between the three-channel plain image and image post-encryption of size 512×512

Image title	Proposed values
Lena	9092
Baboon	8472
Peppers	11181

Table 4 Assessment through Pearson correlation coefficient by plotting scatter diagrams of neighbouring pixels of cipher Baboon and Lena image of various sizes

Images	R-channel Baboon			R-channel Lena		
	I_H	I_V	I_D	I_H	I_V	I_D
Size 256 × 256	0.0010	−0.016	0.04	0.0200	−0.059	0.0050
Size 128 × 128	0.0012	−0.0018	−0.0090	−0.025	−0.0091	0.0126
Size 64 × 64	0.001	−0035	0.008	−0.0086	−0.0062	0.0123
Images	G-channel-Baboon			G-channel-Lena		
Size 256 × 256	−0.0116	0.0199	0.0150	0.0080	0.0150	0.0114
Size 128 × 128	−0.0001	−0.0310	−0.0159	−0.0056	0.0178	−0.0147
Size 64 × 64	−0.0041	0.0005	0.0242	−0.0187	0.0011	0.0123
Images	B-channel-Baboon			B-channel-Lena		
Size 256 × 256	−0.0241	−0.0389	0.0289	0.0083	0.0150	0.0111
Size 128 × 128	−0.0070	0.0230	0.0066	−0.0056	0.0178	−0.0187
Size 64 × 64	−0.0184	−0.0059	0.0084	−0.0187	0.0001	0.016

tion coefficient along these three directions for images of dimension 512 × 512 and 256 × 256. Along this analysis, the following tabulated diagrammatic representation (Figs. 6, 7 and 8) of scatter plots shows channel-wise correlation between cipher and plain three-channel images.

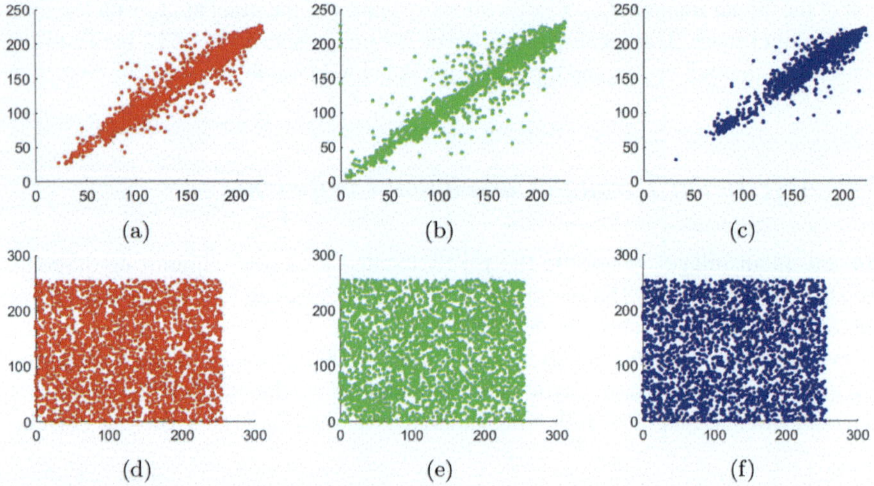

Fig. 6 Horizontal correlation analysis (Baboon) channel-wise **a** R-channel-(plain), **b** G-Channel-(plain), **c** B-channel-(plain), **d** R-channel-(cipher), **e** G-channel-(cipher), **f** B-channel-(cipher)

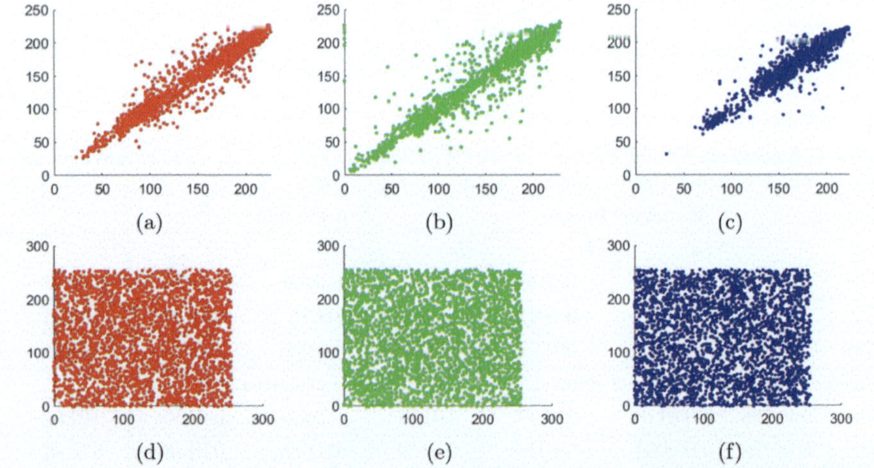

Fig. 7 Vertical correlation analysis (Baboon) channel-wise **a** R-channel-(plain), **b** G-Channel-(plain), **c** B-channel-(plain), **d** R-channel-(cipher), **e** G-channel-(cipher), **f** B-channel-(cipher)

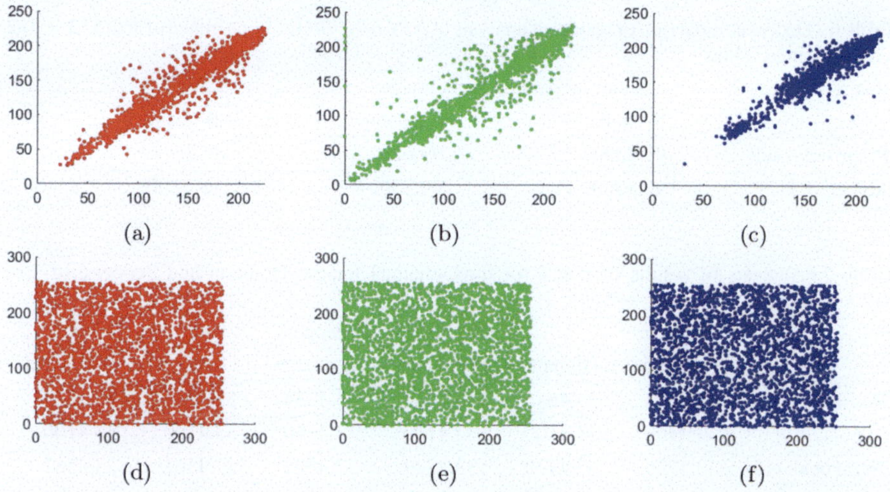

Fig. 8 Diagonal correlation analysis (Baboon) channel-wise **a** R-channel-(plain), **b** G-Channel-(plain), **c** B-channel-(plain), **d** R-channel-(cipher), **e** G-channel-(cipher), **f** B-channel-(cipher)

8 Resistance Against Differential Attack

Differential attacks in cryptographic algorithms are one type of cryptanalysis which are used to understand the effectiveness of encryption schemes against the attacks on block ciphers. The following steps show how differential attack analysis is carried over:

1. I_{ori_1} = plain image, I_{ori_2} = image obtained from I_{ori_1} by minute alteration in one or more pixel positions or intensities.
2. I_{enc_1} = cipher image of I_{ori_1} obtained using proposed scheme of three-channel encryption and I_{enc_2} = cipher image of I_{ori_2} obtained using proposed scheme of three-channel encryption.
3. $\delta I_{enc} = I_{enc_1} \oplus I_{enc_2}$, where $\oplus$ is XOR operation.
4. The analysis of δI_{enc} for understanding differences in encrypted images after alteration and before alteration is referred to as differential attacks.
5. This study involves two major techniques, viz. *NPCR/NBCR*(number of pixels/bits change rate) and *UACI* (unified averaged changed intensity).

Tables 5, 6, 7 and 8 show results of *NPCR/NPBR* and *UACI*.

Table 5 *NPCR/NBPR* and *UACI* analysis results for Baboon, Lena and City images of size 512 × 512

Image title	*NPCR/NPBR*	*UACI*
Lena	99.60930	33.46350
Baboon	99.60900	33.46350
City	99.61901	33.46540

Table 6 Decent comparison of results of analysis of *NPCR/NPBR* for Lena image of size 512 × 512 with reference values

Images	Channel separated Lena image		
	R-channel	G-channel	B-channel
Proposed result	99.5990	99.6202	99.6000
Roy et al. [17]	99.5900	99.5500	99.5800

Table 7 Analysis of results of *UACI* for three-channel images (Baboon and Peppers) of size 512 × 512

	Baboon			Peppers		
	R-channel	G-channel	B-channel	R-channel	G-channel	B-channel
Results	33.4635	33.4635	33.4635	33.4335	33.4335	33.4335
Li et al. [10]	33.4100	33.4100	33.4100	33.4100	33.4100	33.4100
Niyat et al. [12]	33.4729	33.4729	33.4729	33.4562	33.4562	33.4562
Jolfaei et al. [9]	33.4913	33.3786	33.4692			

Table 8 Result analysis of *UACI* values for three-channel images (Baboon and Lena) of size 64 × 64, and 128 × 128 and 256 × 256

Image titles	Size 64 × 64	Size 128 × 128	Size 256 × 256
Baboon	33.3340	34.3300	33.4000
Lena	33.4440	33.4430	33.3301

9 Resistance Against Cutting Channel Attack

To check the strength of any encryption scheme against downfall of information attack, cutting channel analysis is used. If any one channel of a three-channel image is cut 25%, 50% or 75% from the encrypted three-channel image, then the analysis tests how is the recovery of the plain image. Cutting channel attack analysis shows how the encryption scheme withstands against loss of information attack. A piece of Red-channel from the encrypted image is cut 25% from left-top, 25% from bottom-left and 25% from right-top, a total 50%, as well as 75% and retrieved with the help of proposed decryption scheme. The recovery of the cipher images which are cropped in specific percentages is given in Fig. 9.

10 Resistance Against Noise Injection Attacks

The injection of digital distortion in the cipher image alters the original information. In the proposed work, cipher images are distorted by injecting 20% impulse noise and 20% Gaussian noise in cipher images. These distorted images are retrieved using the

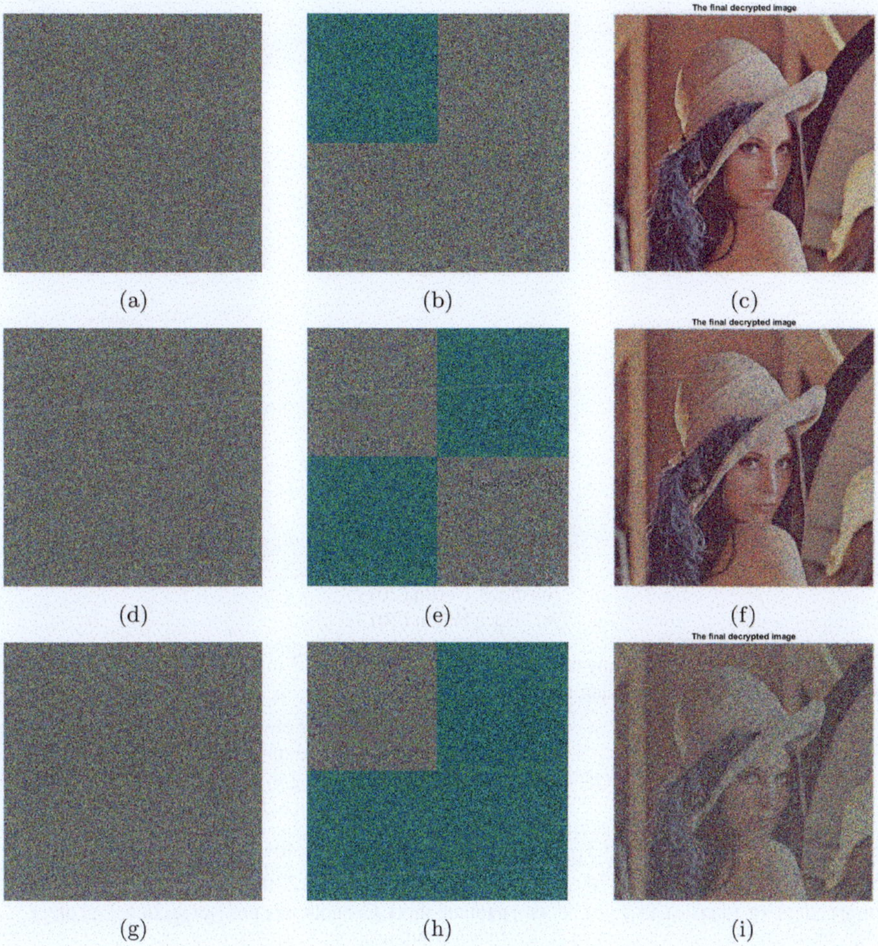

Fig. 9 Lena cropping attack analysis: **a** Encrypted, **b** 25% channel cut attack, **c** Recovered, **d** Encrypted, **e** 50% channel cut attack, **f** Recovered, **g** Encrypted, **h** 75% channel cut attack, **i** Recovered

three-channel encryption scheme proposed in this work. The resultants are tabulated in Fig. 10 which claims a good amount of recovery of original data, hence proving the efficiency against this attack.

10.1 Analysis Through BRISQUE, NIQE, PIQE

Blind/reference less image spatial quality evaluator computes the no-reference image quality ratings for the original image, encrypted image, noisy or cropped image and image after retrieval using the proposed decryption scheme. If the score is less then

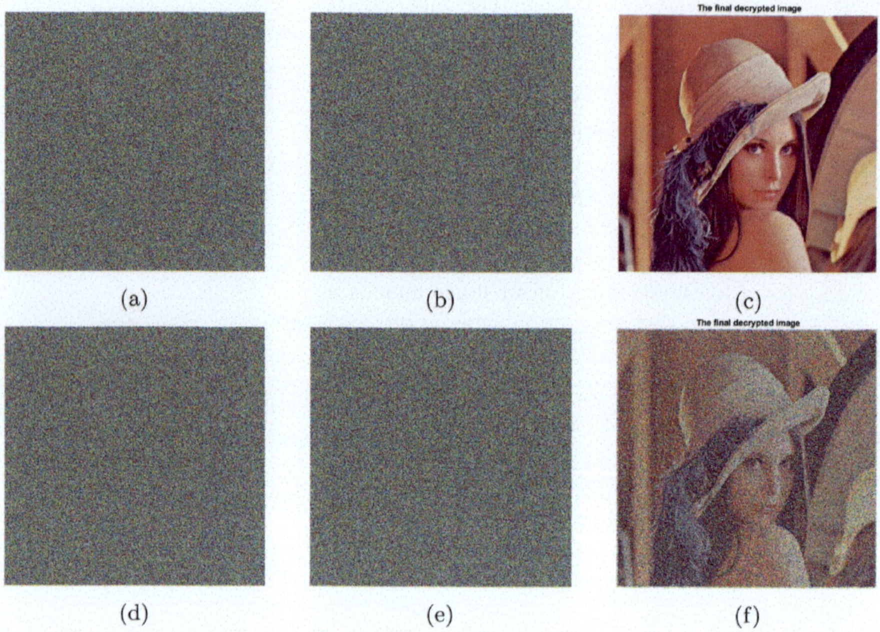

Fig. 10 Lena noise attack analysis: **a** Encrypted, **b** 20% added impulse noise, **c** Recovered, **d** Encrypted, **e** 20% added Gaussian noise, **f** Recovered

Table 9 BRISQUE, NIQE, PIQE scores for Lena image of size 256 × 256

Scores	Original	(i)	(ii)	(iii)	(iv)	(v)	(vi)
BRISQUE	21.4166	29.9263	43.7812	42.9900	43.4042	43.9967	43.9990
NIQE	2.3798	5.9370	10.7486	10.2487	10.3308	11.9104	11.9270
PIQE	32.1841	37.7289	58.1812	63.3148	65.9411	67.8949	67.9023

perpetual quality is better. Also, the naturalness image quality evaluator (NIQE) score evaluates no-reference-quality score for the original image, encrypted image, noisy or cropped image and image after retrieval using the proposed decryption scheme. The lower the score the better the quality. Similarly, to evaluate perception-based image quality, the PIQE score is used. The smaller the score the better the quality. Table 9 shows BRISQUE, NIQE and PIQE scores for the above-mentioned images. Where,

(i) = Original Lena image.

(ii) = 25% cropped and retrieved using decryption scheme.

(iii) = 50% cropped and retrieved using decryption scheme.

(iv) = 75% cropped and retrieved using decryption scheme.

(v) = 20% impulse noise is injected and retrieved using decryption scheme.

(vi) = 20% Gaussian noise is injected and retrieved using a decryption scheme.

Table 10 Statistical analysis using NIST arbitrariness tests for encrypted Baboon image of size 512×512

NIST randomness tests	P-value	Conclusion
Serial test	0.4747	Fulfilled
Block frequency test	0.9838	Fulfilled
Longest run of ones in a block	0.5660	Fulfilled
Binary matrix rank	0.3816	Fulfilled
Frequency test	0.9664	Fulfilled
Approximate entropy	0.2987	Fulfilled
Cumulative sums test	0.8881	Fulfilled
Runs test	0.4894	Fulfilled
Discrete Fourier transform	0.3233	Fulfilled
Non-overlapping template matching	0.2233	Fulfilled
Overlapping template matching	0.2729	Fulfilled
Linear complexity	0.6130	Fulfilled

11 Analysis Through NIST Randomness Tests

The National Institute of Standards and Technology (NIST) tests for checking randomness in samples are a suite of statistical tests used to analyse the arbitrariness of a binary sequence. These tests are designed to assess the quality of randomness in a sequence, which is essential in various fields like simulation, mathematical and statistical modelling along with cryptography and encryption schemes. Verifying the uncertainty in encrypted three-channel images plays a vital role. The most systematized and utilized suite for verification of the existence of the randomness of information post-encryption is the NIST suite. These tests are authentic and very rapid, though they test big multimedia data. The table shows the results of NIST randomness tests for the Baboon cipher image of size 512×512 given in Table 10 and verifies that the results are not biased, strengthening the generation of randomness using the proposed chaotic map.

12 Conclusion

The proposed three-channel image encryption scheme is secure, robust and fast for colour images of varied sizes and varied intensities based on a three-level security mechanism. The scheme uses Hankel transform with geometric transform, Fibonacci $\mathcal{Q}$-Matrix and six-dimensional hyper-chaotic maps as its key features. This scheme incorporates two basic principles of image encryption, viz. confusion and diffusion. The MATLAB 2023a simulations verify that the encryption scheme is fast in terms of computation time required for the execution of the end-to-end scheme. The proposed scheme is tested on multiple attacks like differential attacks, cutting channel attacks and noise attacks, and tabulated results confirm security along with resilience of the proposed three-channel image encryption scheme.

Acknowledgements The authors are thankful to the learned referees for their valuable suggestions to improve the quality of the manuscript.

References

1. Zhu, C., Sun, K.: Cryptanalyzing and improving a novel color image encryption algorithm using RT-enhanced chaotic tent maps. IEEE Access **6**, 18759–18770 (2018)
2. Kumar, M., Gupta, P.: A new medical image encryption algorithm based on the 1D logistic map associated with pseudo-random numbers. Multimed. Tools Appl. **80**(12), 18941–18967 (2021)
3. Wang, H., Tao, X., Huang, J.S.: An improved chessboard covering algorithm with generalized fractal strategy (2019)
4. Cao, W., Cai, H., Hua, Z.: n-dimensional chaotic map with application in secure communication. Chaos, Solitons Fractals **163**, 112519 (2022)
5. Agarwal, S.: Secure image transmission using fractal and 2D-chaotic map. J. Imaging **4**(1), 17 (2018)
6. Hosny, K.M. (ed.): Multimedia security using chaotic maps: principles and methodologies, vol. 884. Springer, Berlin/Heidelberg, Germany (2020)
7. Amin, M., Faragallah, O.S., Abd El-Latif, A.A.: A chaotic block cipher algorithm for image cryptosystems. Commun. Nonlinear Sci. Numer. Simul. **15**(11), 3484–3497 (2010)
8. Dagadu, J.C., Li, J.: Context-based watermarking cum chaotic encryption for medical images in telemedicine applications. Multimed. Tools Appl. **77**, 24289–24312 (2018)
9. Jolfaei, A., Matinfar, A., Mirghadri, A.: Preserving the confidentiality of digital images using a chaotic encryption scheme. Int. J. Electron. Secur. Digit. Forensics **7**(3), 258–277 (2015)
10. Li, C., Luo, G., Qin, K., Li, C.: An image encryption scheme based on chaotic tent map. Nonlinear Dyn. **87**, 127–133 (2017)
11. AlZain, M.A., Faragallah, O.S.: Efficient chaotic tent map-based image cryptosystem. Int. J. Comput. Appl. **167**(7) (2017)
12. Niyat, A.Y., Moattar, M.H., Torshiz, M.N.: Color image encryption based on hybrid hyper-chaotic system and cellular automata. Opt. Lasers Eng. **90**, 225–237 (2017)
13. Elshamy, A.M., Abd El-Samie, F.E., Faragallah, O.S., Elshamy, E.M., El-sayed, H.S., El-Zoghdy, S.F., Alhamad, A.Q.: Optical image cryptosystem using double random phase encoding and Arnold's Cat map. Opt. Quantum Electron. **48**, 1–18 (2016)
14. Mandal, M.K., Banik, G.D., Chattopadhyay, D., Nandi, D.: An image encryption process based on chaotic logistic map. IETE Tech. Rev. **29**(5), 395–404 (2012)
15. Behnia, S., Akhshani, A., Ahadpour, S., Mahmodi, H., Akhavan, A.: A fast chaotic encryption scheme based on piecewise nonlinear chaotic maps. Phys. Lett. A **366**(4–5), 391–396 (2007)
16. Es-Sabry, M., El Akkad, N., Merras, M., Satori, K., El-Shafai, W., Altameem, T., Fouda, M.M.: Securing images using high dimensional chaotic maps and DNA encoding techniques. IEEE Access (2023)
17. Roy, S., Shrivastava, M., Rawat, U., Pandey, C.V., Nayak, S.K.: IESCA: an efficient image encryption scheme using 2-D cellular automata. J. Inf. Secur. Appl. **61**, 102919 (2021)
18. Mohanrasu, S.S., Udhayakumar, K., Priyanka, T.M.C., Gowrisankar, A., Banerjee, S., Rakkiyappan, R.: Event-triggered impulsive controller design for synchronization of delayed chaotic neural networks and its fractal reconstruction: an application to image encryption. Appl. Math. Model. **115**, 490–512 (2023)

19. Gan, Z., Sun, M., Song, Y., Chai, X., Jiang, D., Long, G., He, X.: Visually meaningful image encryption scheme using multi-parameter fractal theory and block synchronous sorting diffusion. Phys. Scr. **98**(8), 085216 (2023)
20. Chang, H., Wang, E., Liu, J.: Research on image encryption based on fractional seed chaos generator and fractal theory. Fractal Fract. **7**(3), 221 (2023)
21. Hosny, K.M., Kamal, S.T., Darwish, M.M., Papakostas, G.A.: New image encryption algorithm using hyperchaotic system and fibonacci q-matrix. Electronics **10**(9), 1066 (2021)
22. Tuncer, T.: The implementation of chaos-based PUF designs in field programmable gate array. Nonlinear Dyn. **86**, 975–986 (2016)

Investigation of Boiler Efficiency Improvements via Enthalpy Wheel with Application to a Biomass Boiler

C. Deniz Canal, M. Diederich, O. Karacay, A. C. Benim, A. Hamberger, M. Heese, and K. H. Schräder

Abstract Based on a mathematical model, the efficiency of a boiler for different utilizations of the exhaust gas energy is systematically investigated. The analysis is carried out for a biomass boiler, which is a novel aspect of the study. It is shown that condensing conditions using a condensing boiler leads to higher efficiencies. Beyond this, it is additionally shown that further efficiency increases can be achieved by the incorporation of a so-called enthalpy wheel, which is a rotary heat and mass exchanging device, even without using the condensation energy. The enthalpy wheel can be operated as a regenerative heat exchanger to preheat the combustion air, which enhances the boiler efficiency. Further, it can additionally serve to moisture recovery, by transferring moisture from the exhaust gas into the combustion air, when its surfaces are covered by a desiccant material. It is shown that the highest efficiencies are obtained when an enthalpy wheel is operated for simultaneous heat and moisture recovery. It is also shown that the exit temperature of the condensing heat exchanger is an important parameter, not only for the case without an enthalpy wheel but also for the enthalpy wheel case, having a substantial influence on the boiler efficiency.

Keywords Condensing boiler · Enthalpy wheel · Biomass boiler

C. Deniz Canal (✉) · M. Diederich · O. Karacay · A. C. Benim
Center of Flow Simulation (CFS), Department of Mechanical and Process Engineering, Düsseldorf University of Applied Sciences, Düsseldorf, Germany
e-mail: cansu.denizcanal@hs-duesseldorf.de

A. Hamberger · M. Heese
Endress Holzfeuerungsanlagen GmbH, Industriestr. 18, 91593 Burgbernheim, Germany

K. H. Schräder
Kamen, Germany

R. K. Sharma et al. (eds.), *Mathematical Theory and Simulation of Scientific Problems*, Springer Proceedings in Mathematics & Statistics 487,
https://doi.org/10.1007/978-981-96-2579-6_13

Abbreviations

Nomenclature

c_p	Molar heat capacity
$\dot{n}$	Molar flow rate
h	Molar enthalpy
$\dot{Q}$	Heat flow rate
T	Temperature
w	Molar water vapor load
w_R	Kmol water vapor produced per kmol fuel
ΔH_{vap}	Molar enthalpy of evaporation
λ	Air factor
η	Boiler efficiency
η_H	Enthalpy efficiency of enthalpy wheel
η_T	Temperature efficiency of enthalpy wheel

Subscripts

a	Air
B	Boiler
EW	Enthalpy wheel
f	Fuel
g	Exhaust gas
S	Saturated state
w	Water

Superscripts

i	Inlet
d	Dry
daf	Dry and ash-free
o	Outlet
v	Vapor

Abbreviations

HHV	Higher heating value
LHV	Lower heating value

1 Introduction

As the world's population continues to grow and economies develop, the demand for energy is also increasing. Industrialization, urbanization, and the spread of energy-intensive technologies have all contributed to an increasing demand for energy. The global energy crisis is a complex and multifaceted problem that poses major challenges for the economy, society, and environment [1]. The efficient use of energy and the use of renewable and clean energy sources are two very important pillars for the energy transition to overcome current and future challenges [2]. The present paper presents addresses the two pillars, namely the efficient energy utilization and renewable energies, where condensation technologies are applied to increase the efficiency of a biomass boiler.

Boilers convert, by combustion, the chemically bound energy of the fuel into internal energy of the exhaust gas, which is then transferred to the water as heat, by cooling down the exhaust gas by water in heat exchangers [3]. Obviously, the useful energy, i.e., the extracted heat from the exhaust gas increases with decreasing exhaust gas exit temperature. The exhaust gas can contain considerable amounts of water vapor due to the combustion of the hydrogen content of the fuel, as well as the humidity of the fuel air streams. The vapor contained in the exhaust gas starts to condensate, when the exhaust gas temperature reaches the dew point temperature [4] as it gets cooled down. Since condensation can cause problems such as corrosion, it is avoided in conventional systems. For this, the exhaust gas exit temperature is to be kept above the dew point temperature. Since the latter lies normally quite above the ambient temperature, a considerable amount of energy potential remains unused.

A remedy to this problem is provided by the condensing heat exchangers that can allow condensation due to their special designs and materials, and thus, recover and utilize the latent heat of condensable matter (water) from exhaust gases [5]. A decrease of the temperature of the cooling medium is beneficial for the operation of the condensing heat exchanger [6, 7]. The condensation process in the heat exchanger can further be supported by increasing the dew point of the hot exhaust gas. A corresponding concept was proposed in the patent of the French Gas Corporation, Gaz de France in 1979 (French patents No. 7900901 and 8,112,770). The inventor R. Guillet named the corresponding device "pompe à vapeur d'eau" that may be translated as "water vapor pump" [8–10]. A thermodynamic analysis of such a system was presented by Kuck [11], who realized the vapor pump with the aid of a spray tower. Further investigations of boilers equipped with different vapor pump systems were provided by Wang et al. [12].

A special type of heat exchanger working with the regenerative principle, i.e., in a transient operation mode is the so-called Ljungström air preheater [13], which is mainly used in thermal power plants to preheat the combustion air, i.e., for heat recovery. Here, the hot gas and cold air flow alternately through the tightly packed numerous narrow channels of the device as the drum rotates. The sensible thermal energy stored in the material of the channel walls during the passage of the hot gas is transferred to the cold air as the cold air flows through the channels, transferring the heat from the exhaust gas to the air.

The so-called enthalpy wheel works with the same regenerative principle with the difference that the channel walls are now coated (or even made out of) desiccant material [14]. By the action of the desiccant material, i.e., adsorption, mass transfer is accomplished between the two streams. This is driven primarily by the difference in the water vapor partial pressures of the streams. Through this mechanism, not only the sensible heat but also the latent heat and moisture is transferred between the streams. The usual area of application of enthalpy wheel, so far, has been the air-to-air HVAC systems [15].

A novel application of the enthalpy wheel in boilers as vapor pump, for increasing the boiler efficiency by increasing the dew point temperature of the exhaust gas, was first presented by Men et al. [16] for a natural gas boiler. The current work follows the same principle for biomass boilers. For the thermodynamic analysis of the system, the approach proposed by Kuck [11] is currently adopted. Thus, the present paper has similarities to the previous publications of Kuck [11] and Men et al. [16].

A difference to the publication of Kuck [11] is given by the fact that in [11] a spray tower was considered as the vapor pump, whereas, in the current study, an enthalpy wheel is considered. A further novelty of the present paper in comparison to the previous publications [11, 16] is given by the fact that a biomass boiler is presently considered, whereas natural gas boilers were considered in the previous studies [11, 16]. In the present paper, the thermodynamic analysis of a condensing biomass boiler utilizing enthalpy wheel as vapor pump is presented.

2 Configuration of Considered Systems

The ingredients of the investigated systems are a boiler, a condensing heat exchanger, and an enthalpy wheel. Although a condensing boiler is envisaged, depending on the cooling efficiency in the heat exchanger, i.e., by the realizability of low-temperature sink, a condensation may or may not take place in the heat exchanger.

In the current analysis, three different configurations are considered, which are defined as follows:

Case A: The boiler is operated in combination with the condensing heat exchanger, but without using the enthalpy wheel.

Case B: The boiler is operated in combination with the condensing heat exchanger and the enthalpy wheel. However, the enthalpy wheel is used only for the transfer of sensible heat (Ljungstrom air preheater), without mass transfer.

Case C: The boiler is operated in combination with the condensing heat exchanger and the enthalpy wheel. In difference to Case B, the full potential of the enthalpy wheel is unfolded, which means that the enthalpy wheel is now equipped with desiccant material. By this means, not only the sensible thermal energy but also the latent energy and moisture are transferred from the exhaust gas to air. The considered systems are sketched in Fig. 1, where the water cycle is not displayed for simplicity.

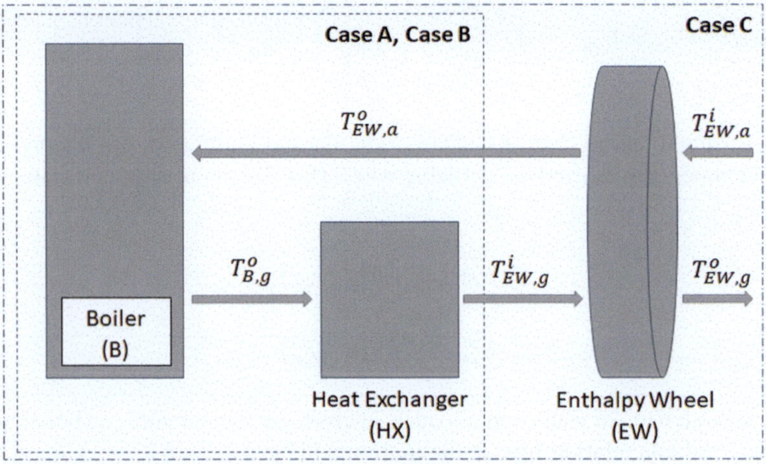

Fig. 1 System configuration

3 Mathematical Modeling

In this section, the mathematical model employed to express the energy balances of the considered systems is presented, which is based on a series of assumptions. In the main line of the mathematical modeling, the approaches suggested by Kuck [11] are adopted. In the mathematical model, a steady-state flow process is assumed, where the changes in the kinetic and potential energies are neglected. Each system component, modeled as lumped, is characterized by the unique inlet and outlet values of the corresponding variables (indicated in Fig. 1). For simplicity, in the energy balance equations, the reference state is not indicated explicitly.

As fuels normally have rather low temperatures, the sensible enthalpy of the fuel is neglected in the energy balances. Below, the energy balance equations for Case A and Cases B and C are presented separately. The system boundaries as well as the corresponding temperatures are illustrated in Fig. 1.

For Case A and Case B, the energy balances without and with condensation are presented in Eqs. (1a) and (1b), respectively:

$$\dot{n}_f \times HHV + \dot{n}_a \times h^{o}_{EW,a} = \dot{Q} + \dot{n}_g \times h^{i}_{EW,g} \tag{1a}$$

$$\dot{n}_f \times HHV + \dot{n}_a \times h^{o}_{EW,a} = \dot{Q} + \dot{n}_g \times \left\{ h^{i}_{EW,g} + \left[w^{o}_{B,g} - w^{i}_{EW,g} \right] \times c_{p,w} \times T^{i}_{EW,g} \right\} \tag{1b}$$

Similarly, for Case C, the energy balance equations without (Eq. 2a) and with condensation (Eq. 2b) are given below:

$$\dot{n}_f \times HHV + \dot{n}_a \times h^{i}_{EW,a} = \dot{Q} + \dot{n}_g \times h^{o}_{EW,g} \tag{2a}$$

$$\dot{n}_f \times HHV + \dot{n}_a \times h^i_{EW,a} = \dot{Q} + \dot{n}_g \times \left\{ h^o_{EW,g} + \left[w^o_{B,g} - w^o_{EW,g} \right] \times c_{p,w} \times T^o_{EW,g} \right\} \tag{2b}$$

It is assumed that in case of condensation, the gas at the exits of the condensing heat exchanger (Eq. 3a) and the enthalpy wheel (Eq. 3b) are in saturated state, which means

$$w^i_{EW,g} = w_s\left(T^i_{EW,g}\right) \tag{3a}$$

$$w^o_{EW,g} = w_s\left(T^o_{EW,g}\right) \tag{3b}$$

The saturated molar water vapor load w_s at given gas temperature can be calculated from the available relationships in the literature [17].

The molar water vapor load at the boiler outlet $w^o_{B,g}$ can be obtained from the following humidity balance:

$$\dot{n}_g \times w^o_{B,g} = \dot{n}_g \times w_R + \dot{n}_a \times w^i_{EW,a} \tag{4}$$

where w_R denotes the water vapor produced by combustion per unit fuel (as kmol H_2O/kmol Exhaust Gas), which depends on the fuel composition. Note that, in Case A and Case B, there is no humidity transfer between the exhaust gas and air (in Case B, the wheel operates purely a as a regenerative heat exchanger), whereas Case C differs from Case B by the transfer of humidity from exhaust gas to air. For different cases, the following expressions are obtained for $w^o_{B,g}$ (for the definition of A in the below equation, please see Eq. (15a).

Case A:

$$w^o_{B,g} = w_R + A \times w^o_{EW,a} \tag{5}$$

Case B:

$$w^o_{B,g} = w_R + A \times w^i_{EW,a} \tag{6}$$

Case C:

$$w^o_{B,g} = w_R + A \times w^o_{EW,a} \tag{7a}$$

where

$$w^o_{EW,a} = w^i_{EW,a} + \left\{ w_s\left(T^i_{EW,g}\right) - w_s\left(T^o_{EW,g}\right) \right\} / A \tag{7b}$$

Further, the variables $h^o_{EW,g}$, $h^i_{EW,a}$ and $h^i_{EW,g}$ $h^o_{EW,a}$ in the above expressions can be calculated from

$$h^o_{EW,g} = c^d_{p,g} \times T^o_{EW,g} + w^o_{EW,g} \times (\Delta H_{vap} + c^v_{p,w} \times T^o_{EW,g}) \tag{8}$$

$$h^i_{EW,g} = c^d_{p,g} \times T^i_{EW,g} + w^i_{EW,g} \times (\Delta H_{vap} + c^v_{p,w} \times T^i_{EW,g}) \tag{9}$$

$$h^o_{EW,a} = c^d_{p,a} \times T^o_{EW,a} + w^o_{EW,a} \times (\Delta H_{vap} + c^v_{p,w} \times T^o_{EW,a}) \tag{10}$$

$$h^i_{EW,a} = c^d_{p,a} \times T^i_{EW,a} + w^i_{EW,a} \times (\Delta H_{vap} + c^v_{p,w} \times T^i_{EW,a}) \tag{11}$$

The boiler efficiency can be defined as

$$\eta = \frac{\dot{Q}}{\dot{n}_f \times HHV} \tag{12}$$

Utilizing the above equations, (Eqns. (1, 2, 12)) expressions for the boiler efficiency can be derived. For Case A and Case B, without (Eq. 13a) and with (Eq. 13b) condensation the following expressions are obtained:

$$\eta = 1 - \frac{E}{HHV}[h^i_{EW,g} - A \times h^o_{EW,a}] \tag{13a}$$

$$\eta = 1 - \frac{E}{HHV}[h^i_{EW,g} + [w^o_{B,g} - w_s\left(T^i_{EW,g}\right)] \times c_{p,w} \times T^i_{EW,g} - A \times h^o_{EW,a}] \tag{13b}$$

Similarly, the following equations can be written for Case C, without (Eq. 14a) and with (Eq. 14b) condensation:

$$\eta = 1 - \frac{E}{HHV}[h^o_{EW,g} - A \times h^i_{EW,a}] \tag{14a}$$

$$\eta = 1 - \frac{E}{HHV}[h^o_{EW,g} + [w^o_{B,g} - w_s\left(T^o_{EW,g}\right)] \times c_{p,w} \times T^o_{EW,g} - A \times h^i_{EW,a}] \tag{14b}$$

In the above equations, the variables A and E are defined as

$$A = \frac{\dot{n}_a}{\dot{n}_g} \tag{15a}$$

$$E = \frac{\dot{n}_g}{\dot{n}_f} \tag{15b}$$

The temperature dependence of the mean molar isobaric heat capacity of air is considered by assuming a third-order polynomial dependence on temperature. The exhaust gas mean isobaric molar heat capacity is calculated based on the exhaust gas composition, calculating the resultant heat capacity for the exhaust gas by a molar weighted sum of the contributions from its components. The heat capacities of the

Table 1 Fuel properties (daf)

HHV MJ/kg	LLV MJ/kg	C % mass	H % mass	O % mass	N % mass
19.91	18.76	47.5	6.6	43.1	2.8

exhaust gas components are, again, calculated assuming a third-order polynomial dependence on temperature [18, 19].

4 Fuel and Air Properties, Boiler Operation Conditions

In the current analysis, the poplar wood is exemplarily considered as the biomass fuel. At the present stage, it is assumed that the sulfur and ash contents of the fuel are negligible. Currently, a mass-based fuel humidity of 25% is assumed. The assumed approximate elementary analysis of the poplar wood as well as the heating values are provided in Table 1 [20, 21].

At the inlet, the relative humidity of air is assumed to be 100% while assuming a temperature of 0 °C. It is assumed that the combustion is complete. The air factor λ is the ratio of the total used amount of air to the theoretical/stoichiometric one. Different values of the air factor between 1 and 2 are considered.

5 Results

5.1 Case A

The calculated boiler efficiencies (Eq. 13b) for three different values of the air factor are displayed in Fig. 2, as functions of the exhaust gas temperature downstream the heat exchanger. This temperature is a parameter that is assumed to be controllable by the corresponding design of the heat exchanger. One can easily recognize that the curves exhibit a kink. This corresponds to the dew point. The dew point temperature is observed to shift to lower values with increasing air factor.

The curve sections for temperatures lower than the dew temperature (curved shapes) display the results with condensation in the heat exchanger (Eq. 13b), whereas, for temperatures larger than the dew point (straight lines) no condensation takes place in the heat exchanger (Eq. 13a).

One can see in Fig. 2 that always higher boiler efficiencies are achieved by decreasing heat exchanger temperatures, i.e., with increased amount of condensation. One can also see that the achievable efficiencies decrease with increasing air factor. However, the differences become smaller with decreasing heat exchanger exit temperatures.

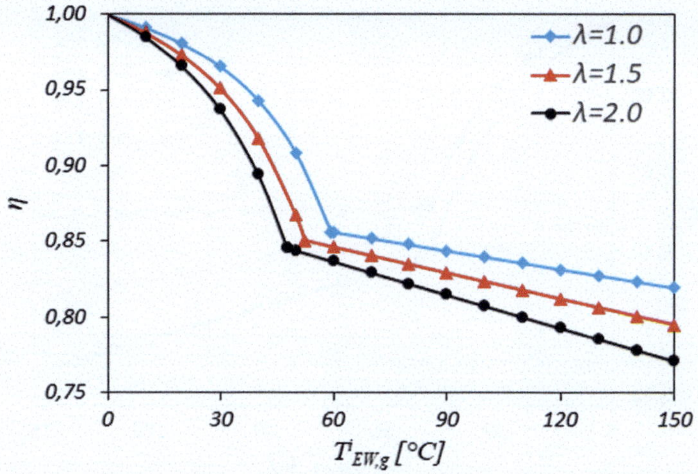

Fig. 2 Efficiency as function of exhaust gas temperature downstream heat exchanger, Case A

5.2 *Case B*

The efficiencies calculated for Case B are displayed in Fig. 3, in a similar manner to Case A (Fig. 2), where the effect of the temperature efficiency of the enthalpy wheel (defined below) is additionally considered. In this case, the enthalpy wheel operates solely as a regenerative heat exchanger (without moisture transfer) having the function of an air preheater. Here, it is important how well the heat transfer is accomplished, i.e., what the temperature efficiency of the heat exchanger is. The temperature efficiency of the enthalpy wheel can be defined as

$$\eta_T = \frac{T^o_{EW,a} - T^i_{EW,a}}{T^i_{EW,g} - T^i_{EW,a}} \tag{16}$$

The efficiency complements the equation system (Eq. 1a) in such a way that the temperature (enthalpy) at the air outlet of the enthalpy wheel $T^o_{EW,a}$ is related to the air and exhaust gas inlet and temperatures of the enthalpy wheel.

At this stage, we don't attempt to calculate/estimate any temperature efficiency for the heat exchanger, but obtain results for different values of this parameter. The realizability of any value is the subject of the design of the particular device, which is not yet attempted at this point.

The curves shown in Fig. 3 for zero temperature efficiency correspond to the curves that were already obtained without enthalpy wheel, i.e., for Case A (Fig. 1). With air preheat the dew point temperature does not change, but the efficiency increases. One

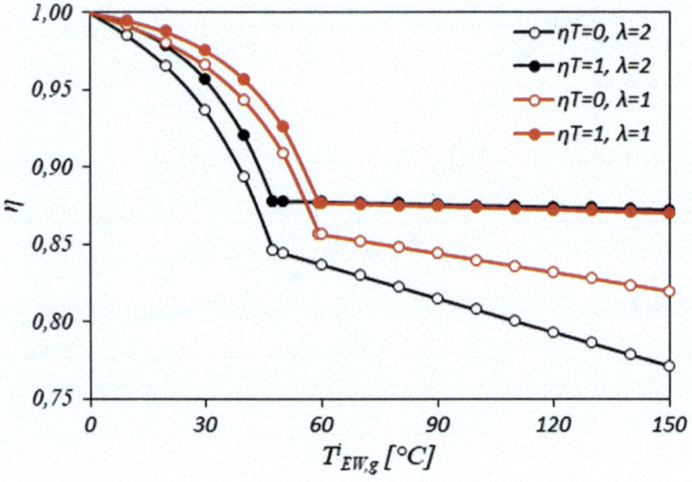

Fig. 3 Efficiency as function of exhaust gas temperature downstream heat exchanger, Case B

can see that with a well-designed regenerative heat exchanger (enthalpy wheel), operating with high temperature efficiency, considerable increases in the boiler efficiency can be achieved.

5.3 Case C

For the calculation of the efficiency of Case C, Eq. (14) is used. Here, in difference to Case B, not only heat but also moisture transferred from exhaust gas to the air, by virtue of the desiccant layers applied to the enthalpy wheel channel surfaces. Here, the performance of the enthalpy wheel, cannot fully be described by the temperature efficiency. Instead, an enthalpy efficiency is defined as follows:

$$\eta_H = \frac{h^i_{EW,g} - h^o_{EW,g}}{h^i_{EW,g} - h^i_{EW,a}} \tag{17}$$

which also accounts for the moisture transport.

The calculated boiler efficiencies for different values of the enthalpy efficiency are displayed in Fig. 4 as function of $T^i_{EW,g}$, for an air factor of 2. Comparing to the previous cases (Case A, Fig. 2, Case B, Fig. 3), one can see that much higher efficiencies can be achieved, depending on the enthalpy efficiency of the wheel (Fig. 4).

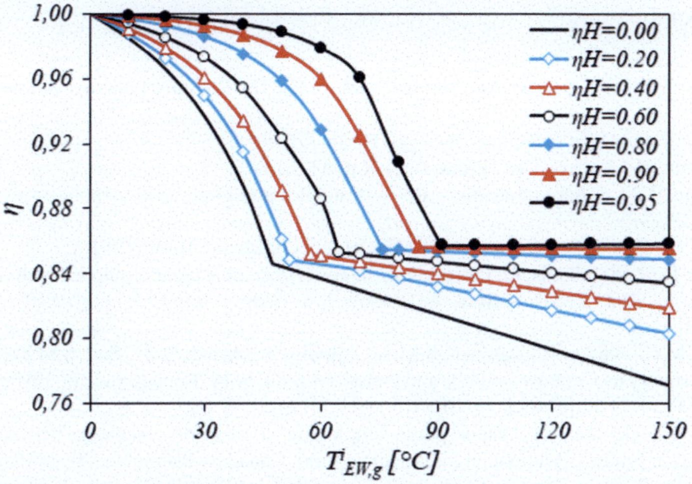

Fig. 4 Efficiency as function of exhaust gas temperature downstream heat exchanger, Case C

6 Conclusions

Applying a mathematical model for the energy balances of a wood burning biomass boiler, are investigated under certain assumptions. According to the model, it is observed that condensing conditions using a condensing boiler enhance the boiler efficiency. Moreover, it is also observed that a further increase can be achieved by the incorporation of a so-called enthalpy wheel, which is a rotary heat and mass exchanging device, even without using the condensation energy, which can also be operated as a regenerative heat exchanger to preheat the combustion air enhances the boiler efficiency. The effect of moisture transfer from the exhaust gas to the combustion air via the dessicant material is further enhanced when the surfaces are coated with a dessicant.

It is shown that the highest efficiencies are obtained when an enthalpy wheel is operated for simultaneous heat and moisture recovery.

Acknowledgements The work was accomplished within the framework of the project "BiHsMod" in the Energy Research Program funded by the German Federal Ministry of Economic Affairs and Climate Protection. Mr. Jan H. Kramb and Mr. Klaus Schmitt of K. Schräder Nachf., Germany, are acknowledged for their valuable technical support. We also gratefully thank Prof. Dr.-Ing. Jürgen Kuck of Ostfalia University of Applied Sciences, Germany, for his valuable support in the theoretical elaboration.

References

1. Von der Gracht, H., Salcher, M., Kerssenbrock, N.G.: The Energy Challenge. Redline, Munich (2016)
2. Petit, V.: The energy transition. Springer, Berlin (2017)
3. Rayaprolu, K.: Boilers. CRC Press, Boca Raton (2013)
4. Whitman, A.M.: Thermodynamics: Basic Principles and Engineering Applications. Springer, Berlin (2023)
5. Bart, H.-J., Scholl, S.: Innovative Heat Exchangers. Springer, Berlin (2018)
6. Lazarin, R.M., Schibuola, L.: Performance analysis of heating plants equipped with condensing boilers. J. Heat Recover. Syst. **6**(4), 269–276 (1986). https://doi.org/10.1016/0198-7593(86)90112-8
7. Barma, M.C., Saidur, R., Rahman, S.M.A., Allouhi, A., Akash, B.A., Sait, S.M.: A review on boilers energy use, energy savings, and emissions reductions. Renew. Sustain. Energy Rev. **79**, 970–983 (2017). https://doi.org/10.1016/j.rser.2017.05.187
8. Guillet, R., Georges, B.: Vapor pump employing counterflow exchange between air and combustion products without an intermediate fluid. Canadian Patent No. 2028485 (1989)
9. Guillet, R.: Pompe à vapeur et générateur à condensation. Revue Générale de Thermique. **30**, 388–393 (1991)
10. Guillet, R.: Wasserdampfpumpe und erdgasbefeuerte Maschinen. Ki Klima-Kälte-Heizung. **21**, 406–409 (1993)
11. Kuck, J.: Efficiency of vapour-pump-equipped condensing boilers. Appl. Therm. Eng. **16**(3), 233–244 (1995). https://doi.org/10.1016/1359-4311(95)00067-4
12. Wang, J., Hua, J., Fu, L., Wang, Z., Zhang, S.: A theoretical fundamental investigation on boilers equipped with vapor-pump system for flue-gas heat and moisture recovery. Energy. **171**, 956–970 (2019). https://doi.org/10.1016/j.energy.2019.01.062
13. Vulloju, S., Kumar, E.M., Kumar, M.S., Reddy, K.K.: Analysis of performance of Ljungstrom air preheater elements. Int. J. Curr. Eng. Technol. **2**, 501–505 (2014). https://doi.org/10.14741/ijcet/spl.2.2014.94
14. El-Maghalany, W.E., ElHefni, A.A., ElHelw, M., Attia, A.: Novel air conditioning system configuration combining sensible and desiccant enthalpy wheels. Appl. Therm. Energy. **127**, 1–15 (2017). https://doi.org/10.1016/j.applthermaleng.2017.08.020
15. De Antonellis, S., Intini, M., Joppolo, C.M., Leone, C.: Desing optimization of heat wheels for energy recovery in HVAC systems. Energies. **7**(11), 7348–7367 (2014). https://doi.org/10.3390/en7117348
16. Men, Y., Liu, X., Zhang, T., Xu, X., Jiang, Y.: Novel flue gas waste heat recover system equipped with enthalpy wheel. Energy Convers. Manage. **196**, 649–663 (2019). https://doi.org/10.1016/j.enconman.2019.06.026
17. Huang, J.: A simple accurate formula for calculating saturation vapor pressure of water and ice. J. Appl. Meteorol. Climatol. **57**(6), 1265–1272 (2018). https://doi.org/10.1175/JAMC-D-17-0334.1
18. Baehr, H.D., Kabelac, S.: Thermodynamik–Grundlagen und technische Anwendungen, 16th edn. Springer, Berlin (2013)
19. Baehr, H.D., Stephan, K.: Wärme-und Stoffübertragung, 10th edn. Springer, Berlin (2019)
20. Watter, A.: Regenerative Energiesysteme. Springer, Berlin (2022)
21. Epple, B., Leithner, R., Linzer, W., Walter, H. (eds.): Simulation von Kraftwerken und Feuerungen, 2nd edn. Springer, Berlin (2012)

To Study the Effect of ER Flux and Orai Flux on Fractional Calcium Diffusion in Neuron Cell

Vora Hardagna Vatsal, Brajesh Kumar Jha, and Tajinder Pal Singh

Abstract In the nervous system, there are many mechanisms to respond to cognitive functions. Neurons are basic cells to perform this cognitive response. Cellular homeostasis of the free calcium ions is governed by the membrane channels and receptors. For performing this approach, we have modeled the neuronal calcium homeostasis for the Caputo fractional time diffusion with the buffer reaction with endoplasmic reticulum flux. Various neuronal organelles like endoplasmic reticulum (ER), Sarcoendoplasmic Reticulum Calcium ATPase (SERCA), inositol triphosphate (IP3) receptor (IP3R) and ryanodine receptor (RYR), and perform their vital role for cellular processes in a calcium dependent manner. The fractional reaction–diffusion model is solved by the Laplace and Fourier integral transform approach. For the closed form Mittag Leffler type solution followed by Green's function. Our obtained solution is in an analytical approach with closed form in Mainardi's function and Wright's function. This simulation shows the significant impacts of different parameters on temporal and spatial diffusion. Different buffer concentrations and different temporal fractional order better with real life physiological properties to mimic the neuronal impact. Our results enrich the significant study of the buffers and calcium relation.

Keywords Calcium distribution · ER fluxes · Buffer · Laplace transforms · Fractional derivative

---

V. H. Vatsal (✉) · B. K. Jha · T. P. Singh
Department of Mathematics, School of Technology, Pandit Deendayal Energy University, Raysan, Gandhinagar, Gujarat-382426, India
e-mail: hardagna.vora@gmail.com

B. K. Jha
e-mail: brajeshjha2881@gmail.com

T. P. Singh
e-mail: tajinder.singh@spt.pdpu.ac.in

R. K. Sharma et al. (eds.), *Mathematical Theory and Simulation of Scientific Problems*, Springer Proceedings in Mathematics & Statistics 487,
https://doi.org/10.1007/978-981-96-2579-6_14

1 Introduction

Calcium acts as the second messenger in the process of cellular transmission. This transmission is responsible for the cellular function of people in many cell lines. Theoretical development involves mimicking calcium oscillations for neural study. This analysis enhances the comprehension of the underlying factors that may affect the calcium flux, such as buffer, diffusion coefficient, various channel mechanisms, and so on.

The concentration of calcium has a crucial role in facilitating the connection between neuroglia cells, promoting neuron growth, moderating apoptosis, and accomplishing other essential actions that are vital for neurons and have a significant impact on governing the human body. Disrupted calcium activity may impair neural processes critical for the efficient operation of human cognitive abilities.

An excessive amount of calcium could potentially cause the neural circuit to explode. Neurodegeneration arises from the dysfunction of organelles caused by an imbalance in calcium levels. The deregulation of neurons may be the cause of Alzheimer's and other neurodegenerative disorders. The symptoms of this condition initially manifest as cognitive impairments, such as memory loss, illusions, and difficulties in forming new memories.

The process of calcium diffusion encompasses numerous intricate mechanisms that are essential for cellular functioning. The essence of the problem can be understood via the lens of the generalized reaction–diffusion equation. The anomalous process is examined using fractional differential analysis, which provides the most accurate fit to real-life events.

In recent years, experts have approached the examination of the calcium hypothesis from many approaches. GD Smith initiated this calcium trail to investigate the effects of calcium buffering [1]. Parotid cells exhibit calcium waves, which are studied using a mathematical technique [2]. Smith GD et al. conducted an analysis of the calcium diffusion process, specifically examining the influence of buffer and source on a prolonged time scale [3]. Astrocytes and neurons, which are distinct types of brain cells, are also simulated to study the effects of calcium distribution. Tewari and Majumdar conducted a simulation of astrocyte cell and synaptic plasticity to study the activity of glutamate receptors [4]. Manhas et al. conducted a study on the intricate calcium waves in pancreatic cells [5]. The study conducted by Jha A et al. utilized a finite element approach to investigate the sodium calcium exchange mechanism in neurons [6]. Dave and Jha combined calcium diffusion with the study of neurodegenerative illness [7]. Dave and Jha have documented the effects of voltage-gated calcium channels on neuronal cells [8]. The study utilized the triangular finite element method to analyze the calcium distribution in multipolar neuronal cells [9]. Mainardi and Pagnini [10] investigated the generalized diffusion technique. Joshi H. et al. have depicted the fractional impact of calcium concentration on nerve cells [11]. The study conducted by Joshi and Jha [12] explored the integration of Parkinson's disease with a fractional calcium reaction diffusion method. Joshi and Jha conducted a two-dimensional analysis of mitochondrial flow [13]. The study investigated a fractional temporal technique to analyze the calcium advection diffusion in astrocytes [14]. The mechanism of Stim Orai is investigated using the fractional time diffusion equation [15]. Jethanandani et al. have studied the neuron and astrocyte dynamics [16]. Joshi and Jha analyzed Parkinson's disease by studying calcium dynamics [17]. Jha et al. have studied excess buffer and calcium dynamics in astrocytes with finite element method in two dimensions [18]. Jethanandani et al. have studied the bifurcation dynamics in neuron with calcium homeostasis [19]. Vora et al. studied the Orai flux dynamics on neurons for calcium distribution [20–22]. Pawar and Pardasani studied the neuronal calcium dynamics using fractional approach [23]. Further they have also studied the nonlinear dynamics of nitric oxide impact on cells [24].

Based on the findings of this literature analysis, we can now present our model. Our model incorporates the time fractional reaction–diffusion technique, taking into account a leak, SERCA channel, and Orai channel flow at the cell boundary. This particular combination of components has not been previously examined.

2 Model Formulation

Calcium follows the reaction diffusion law of mass action and Fick's law. From the natural cellular environment, we have developed the mathematical model with buffer and the written [25],

$$\left[Ca^{+2}\right] + [B] \underset{k^-}{\overset{k^+}{\Longleftrightarrow}} [CaB] \tag{1}$$

By using this fundamental law, we have

$$\frac{\partial\left[Ca^{+2}\right]}{\partial t} = D_{Ca}\nabla^2\left[Ca^{+2}\right] - k^+[B]\left[Ca^{+2}\right] + k^-[CaB] - J_{SERCA} + J_{IPR} + J_{RYR}$$

And boundary condition is [26],

$$\frac{dCa^{+2}(\pm\infty, t)}{dx} = J_{ORAI} \tag{2}$$

where is $D_{Ca} > 0, -\infty < x < \infty, 0 \leq t < \infty$

Where k^+ is the association rate and k^- is the disassociation rate, B is buffer concentration which makes bond to calcium-free ions. D_{Ca} is the calcium diffusion coefficient. J_{SERCA} is the calcium extraction mechanism of the endoplasmic reticulum, and J_{Leak} is leak of calcium ion from endoplasmic reticulum to the cytosol, J_{chan} is channel flux of calcium to the soma, which formulation given below as [16],

$$J_{SERCA} = V_{SERCA}\frac{C^2}{K_{SERCA}^2 + C^2} \tag{3}$$

RyR and IPR and Orai is formulated by,

$$J_{RyR} = k_{RyR}P_{RyR}\left(\left[Ca^{2+}\right]_{ER} - \left[Ca^{2+}\right]\right) \tag{4}$$

$$J_{IPR} = k_{IPR}O_{IPR}\left(\left[Ca^{2+}\right]_{ER} - \left[Ca^{2+}\right]\right) \tag{5}$$

$$J_{ORAI} = \phi\frac{I_{\mathrm{ORAI}}}{A_O * z * F} \tag{6}$$

where, V_{SERCA} is Pump conductance, K_{SERCA} Dissociation pump rate, k_{IPR} Association rate of IPR, O_{IPR} is the opening probability of the IPR channel is between 0 to 1, A_O is area of channel, z is valency of calcium ion, Fis faraday constant, $\left[Ca^{2+}\right]$ is cytosolic calcium concentration. Transforming the integer order of time differential term to non-integer term where $0 < u \leq 1$, [27]

$${}_0^C D_t^u \mathrm{C} = \mathrm{D}\frac{\partial^2 C}{\partial x^2} - aC + b \tag{7}$$

Initial calcium concentration is taken as,

$$C(x, 0) = g(x).$$

3 Solution

To solve this problem, we will use integral transform method and Green's function.

Applying the Fourier cosine transform on pace variable and Laplace transform on time for given physiological parameters with initial and boundary condition [28].

$$\tilde{\widehat{C}}\ (k, \mathrm{s}) = \frac{\mathrm{s}^{\mathrm{u}-1}\widehat{g}(k)}{\mathrm{s}^{\mathrm{u}} - \mathrm{D}k^2 + a} + \frac{(b + j_o)}{s(\mathrm{s}^{\mathrm{u}} - \mathrm{D}k^2 + a)}\delta(k) \tag{8}$$

where k is Fourier variable and s is Laplace variable and applying the inverse transform to above equation using inverse transformation equation [28].

$$E_u\left(pt^u\right) \leftarrow L \rightarrow \frac{s^{u-1}}{s^u - p}\&$$

$$t^{\gamma-1}E_{\mu,\gamma}(\pm pt^{\mu}) \leftarrow L \rightarrow \frac{s^{\mu-1}}{s^{\mu} \mp p} \tag{9}$$

$$C(x,t) = \frac{1}{2\pi}\int_0^{\infty} coskxE_u\left[\{Dk^2 - a\}t^u\right]\hat{g}(k)dk + (b + j_o)t^u$$
$$\frac{1}{2\pi}\int_0^{\infty} coskx\delta(k)E_{u,u+1}\left[\{Dk^2 - a\}t^u\right]dk \tag{10}$$

From this using the greens function for fundamental solution of the fractional reaction diffusion problem,

$G(x,t) = \frac{\sqrt{2}}{\sqrt{\pi}}\int_0^{\infty} coskx\mathrm{E_u}\left[\{Dk^2 - a\}\mathrm{t^u}\right]dk$

$$+(b + j_o)t^u\frac{\sqrt{2}}{\sqrt{\pi}}\int_0^{\infty} coskx\mathrm{E_{u,u+1}}\left[\{Dk^2 - a\}\mathrm{t^u}\right]dk \tag{11}$$

To solve Green's function,

we have applied the Laplace transform and Fourier cosine transform [29, 30],

$L\{\phi(-\mu, 1; -pt^{-\mu})\} = s^{-1}e^{-ps^{\mu}}, -1 < \mu < 0, p > 0$

$L\{t^{-\mu}M_{\mu}\left(\frac{p}{t^{\mu}}\right)\} = s^{\mu-1}e^{-ps^{\mu}}, 0 < \mu < 1, p > 0$

Then we get,

$$G(x,t) = \frac{\sqrt{\pi}}{\sqrt{2Dt^u}}\int_0^{\infty} e^{-\frac{x^2}{4t^u k} - akt^u}k^{-\frac{1}{2}}M_u(k)dk$$
$$+\frac{(b + j_o)t^u\sqrt{\pi}}{\sqrt{2Dt^u}}\int_0^{\infty} e^{-\frac{x^2}{4t^u k} - akt^u}k^{-\frac{1}{2}}\Phi(-u, 1; -k)dk \tag{12}$$

where $M_u(k)$ is the Mainardi's function and $\phi(u, v; k)$ is wright function [31].

4 Result and Discussion

The results are simulated using MATLAB for graphical analysis. The simulation approach is employed to visually represent the spatial and temporal pattern of calcium oscillation. The values of the necessary parameters are provided in Table 1. We can replicate the conditions of the neuron for both a healthy neuron and one damaged by Alzheimer's disease.

The temporal distribution of the calcium profile is depicted in Fig. 1. The calcium profile exhibits an early surge, reaching a maximum of 1.1 μM. Following a rapid movement, it scatters within the cytosolic environment and attaches to a particular substance before being taken in by organelles such as the endoplasmic reticulum. However, the non-integer order transformation demonstrates a raise in calcium ion levels from 0.8 to 1. Furthermore, over a period of time, all lines converge.

Table 1 Physiological values for computation

Notation	Parameter	Value	Unit
D_{Ca}	Diffusion coefficient	150–250	$\mu m^2/s$
C_1	Cell ratio	0.185	
k^-	Association rate of EGTA	1.5	$\mu M^{-1}s^{-1}$
k^-	Association rate of calbindin and calmodulin	75–120	$\mu M^{-1}s^{-1}$
$[Ca^{+2}]_\infty$	Background Calcium concentration	0.1	μM
V_{SERCA}	SERCA pump conductance	120	$s^{-1}(\mu M)^{-2}$
K_{IPR}	IPR association rate	0.52	s^{-1}
A_o	Area of Orai channel	0.25	nm^2
[B]	Buffer concentration	50–100	μM
I_o	Current of Orai channel	2.1	fA

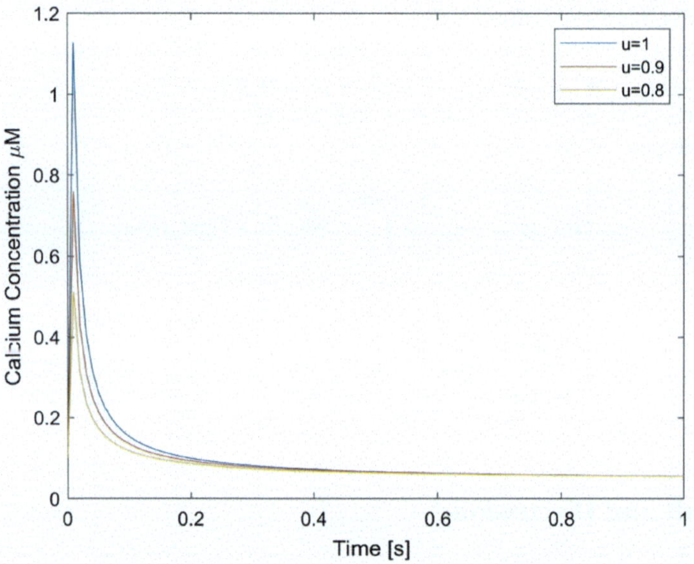

Fig. 1 Calcium concentration against the time with different temporal order

The spatial distribution of calcium concentration is uniformly distributed from the boundary source into the neuron soma, as shown in Fig. 2. The behavior of the cytosolic calcium concentration aligns with the previously observed findings for fractional order u = 0.9, 0.8, and integer order u = 1 [28]. The cellular membrane and organelles exhibit calcium effluxes with a peak concentration of approximately 1 μM. Varying the sequence of time differentiation had a notable effect on calcium concentration, however the curvature of calcium distribution remained consistent.

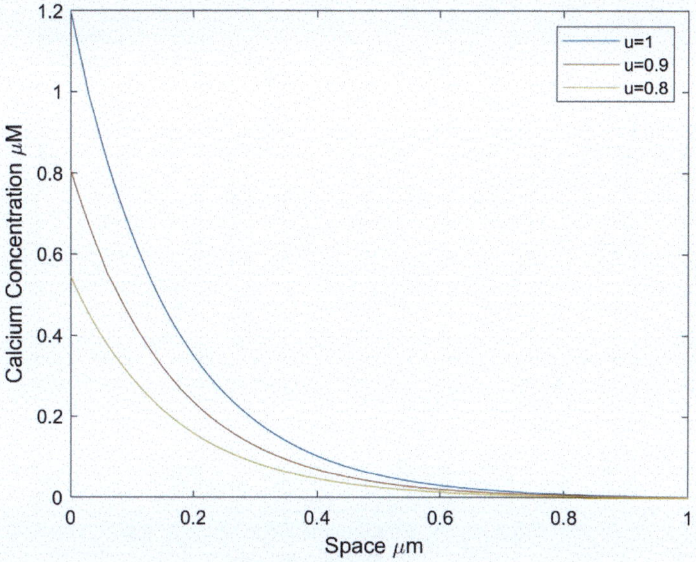

Fig. 2 Calcium concentration against the space with different fractional order

By referring to Fig. 3, we can observe the impact of various buffers on the calcium concentration in neurons. Calmodulin and calbindin play a crucial role as buffers in neurodegenerative diseases [32]. These functions are particular to various neurons. Their role is neuroprotective, which is observable. These two buffers are regulating the calcium content. This implication can result in cellular malfunction and an excessive disruption of calcium homeostasis, ultimately leading to the demise of neuron cells.

Figure 4 displays the spatial distribution of cytosolic calcium ion concentration $[Ca^{2+}]$ under normal and Alzheimer's conditions. The buffer concentration is set at B = 50 μM and B = 100 μM, with an association rate of $k_+ = 1.5\mu M^{-1}s^{-1}$. An observable correlation is seen between the concentration of buffer and the profile of cytosolic calcium. A higher calcium concentration at B = 50 μM or below may indicate the advancement of a neural disease.

The spatial distribution of cytosolic calcium ion concentration $[Ca^{2+}]$ is depicted in Fig. 5, with varying values of the diffusion coefficient. Specifically, the diffusion coefficients of $250\mu m^2 s^{-1}$ and $100\mu m^2 s^{-1}$ are considered, which have implications for the calcium concentration level. Reducing the transport of calcium ions results in the accumulation of calcium near the membrane, which in turn affects cellular activities. By examining the graphical depiction, we may identify many parameters that can be incorporated to demonstrate a major influence. Here we can witness the influence of buffer on the SERCA channel, as well as the leak and boundary flux of Orai. Based on the simulation results, these findings are consistent with earlier results.

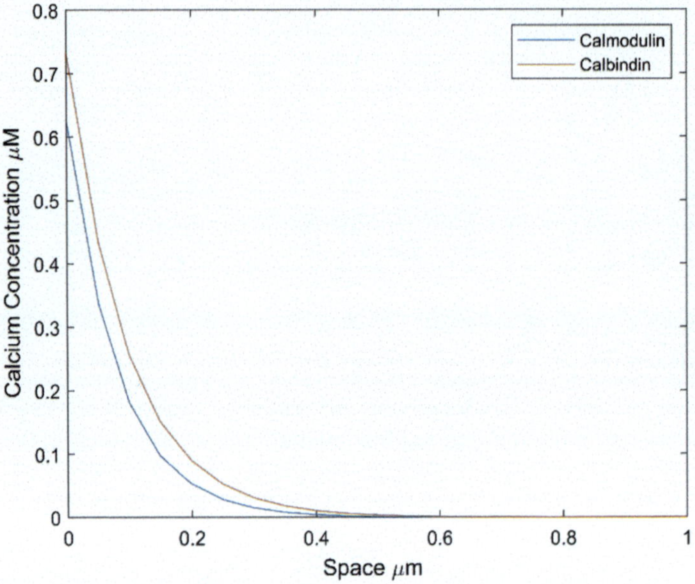

Fig. 3 Calcium concentration against the space with different buffer Calbindin and Calmodulin

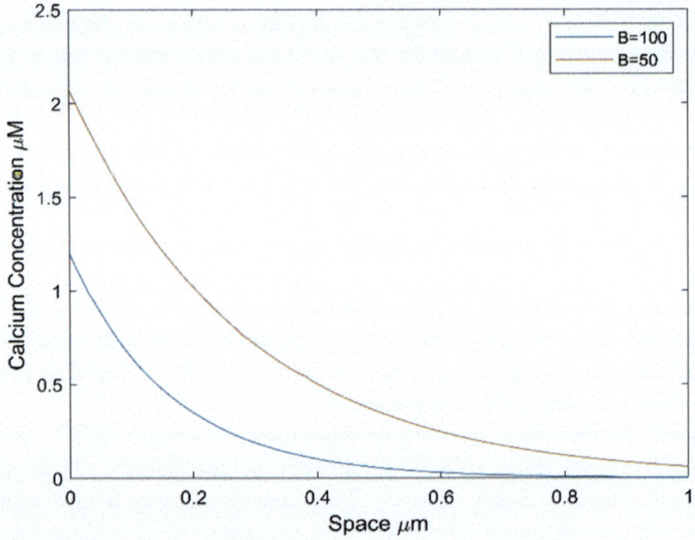

Fig. 4 Calcium concentration against the space with different buffer value

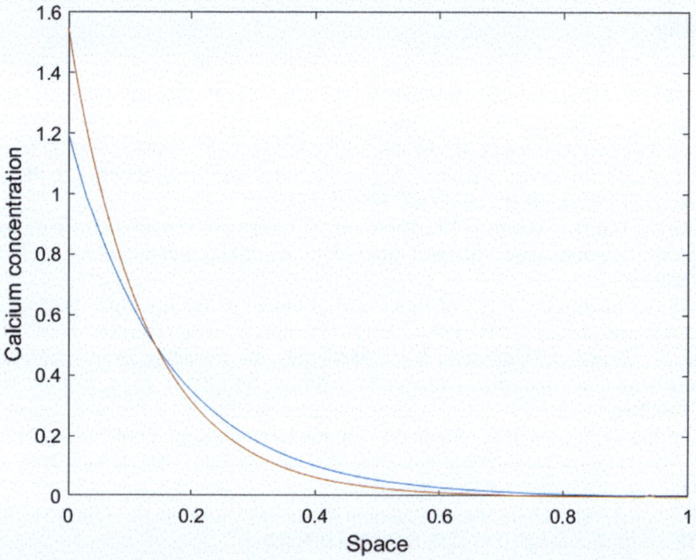

Fig. 5 Calcium concentration against the space with D = 100 (red line), and 250 (blue line)

5 Conclusion

We have constructed the ER model incorporating store-operated calcium entry with boundary flux from the Orai channel. The closed form solutions of several types of Mittag–Leffler family functions, Mainardi's function, and Wright function were derived through the use of Green's function.

The method incorporates the Laplace and Fourier cosine transform. Visualizing the closed form solution showed the significant impact of ER and Orai flux and implication on neuron. Spatio-temporal calcium distribution shows the nature of calcium concentration with different non-integer order transformation. Then changing the diffusion coefficient also elevated the calcium concentration and accumulation showed around the nerve cell membrane.

Alzheimer impacted neuron observed with low buffer and this simulation shows this impact on calcium concentration. Buffer decreased and calcium overload to cell death. This cell loss led to dementia. Using fractional memory approach enhance the behavior and controlling strategy. Which develops other medicine combination to consider for biologist.

References

1. Smith, G.D.: Analytical steady-state solution to the rapid buffering approximation near an open Ca2+ channel. Biophys J. **71** (1996). https://doi.org/10.1016/S0006-3495(96)79500-0
2. Sneyd, J., Tsaneva-Atanasova, K., Bruce, J.I.E., Straub, S.V., Giovannucci, D.R., Yule, D.I.: A model of calcium waves in pancreatic and parotid acinar cells. Biophys. J. **85**, 1392–1405 (2003). https://doi.org/10.1016/S0006-3495(03)74572-X
3. Smith, G.D., Dai, L., Miura, R.M., Sherman, A.: Asymptotic analysis of buffered calcium diffusion near a point source. SIAM J Appl Math. **61** (2001). https://doi.org/10.1137/S0036139900368996
4. Tewari, S.G., Majumdar, K.K.: A mathematical model of the tripartite synapse: Astrocyte-induced synaptic plasticity. J Biol Phys. **38** (2012). https://doi.org/10.1007/s10867-012-9267-7
5. Manhas, N., Sneyd, J., Pardasani, K.R.: Modelling the transition from simple to complex Ca2+oscillations in pancreatic acinar cells. J Biosci. **39** (2014). https://doi.org/10.1007/s12038-014-9430-3
6. Jha, A., Adlakha, N., Jha, B.K.: Finite element model to study effect of Na+-Ca2+ exchangers and source geometry on calcium dynamics in a neuron cell. J Mech Med Biol. **16** (2016). https://doi.org/10.1142/S0219519416500184
7. Dave, D.D., Jha, B.K.: Delineation of calcium diffusion in alzheimeric brain. J Mech Med Biol. **18** (2018). https://doi.org/10.1142/S0219519418500288
8. Dave, D.D., Jha, B.K.: Modeling the alterations in calcium homeostasis in the presence of protein and VGCC for alzheimeric cell. In: Advances in Intelligent Systems and Computing (2018)
9. Dave, D.D., Jha, B.K.: On finite element estimation of calcium advection diffusion in a multipolar neuron. J Eng Math. **128** (2021). https://doi.org/10.1007/s10665-021-10129-z
10. Mainardi, F., Pagnini, G.: The fundamental solutions of the time-fractional diffusion equation. Presented at the (2002)
11. Jha, B.K., Joshi, H., Dave, D.D.: Portraying the effect of calcium-binding proteins on cytosolic calcium concentration distribution fractionally in nerve cells. Interdiscip Sci. **10** (2018). https://doi.org/10.1007/s12539-016-0202-7
12. Joshi, H., Jha, B.K.: Fractional-order mathematical model for calcium distribution in nerve cells. Comput. Appl. Math. **39** (2020). https://doi.org/10.1007/s40314-020-1082-3
13. Joshi, H., Jha, B.K.: 2D memory-based mathematical analysis for the combined impact of calcium influx and efflux on nerve cells. Comput. Math. Appl. **134**, 33–44 (2023). https://doi.org/10.1016/j.camwa.2022.12.016
14. Jha, B.K., Vatsal, V.H., Joshi, H.: A fractional approach to study of calcium advection distribution and VGCC in Astrocyte. In: 2023 International Conference on Fractional Differentiation and Its Applications (ICFDA), pp. 1–5. IEEE (2023)
15. Vatsal, V.H., Jha, B.K., Singh, T.P.: To study the effect of ER flux with buffer on the neuronal calcium. Eur. Phys. J. Plus. **138**, 494 (2023). https://doi.org/10.1140/epjp/s13360-023-04077-z
16. Jethanandani, H., Jha, B.K., Ubale, M.: The role of calcium dynamics with amyloid beta on neuron-astrocyte coupling. Math. Model. Numer. Simul. Appl. **3**, 376–390 (2023). https://doi.org/10.53391/mmnsa.1398320
17. Joshi, H., Jha, B.K.: Chaos of calcium diffusion in Parkinson's infectious disease model and treatment mechanism via Hilfer fractional derivative. Math. Model. Numer. Simul. Appl. **1** (2021). https://doi.org/10.53391/mmnsa.2021.01.008
18. Jha, B.K., Adlakha, N., Mehta, M.N.: Two-dimensional finite element model to study calcium distribution in astrocytes in presence of excess buffer. Int. J. Biomath. **7** (2014). https://doi.org/10.1142/S1793524514500314
19. Jethanandani, H., Jha, B.K., Ubale, M.: Bifurcation analysis of calcium dynamics in nerve cell. Eur. Phys. J. Plus. **138**, 1159 (2023). https://doi.org/10.1140/epjp/s13360-023-04699-3
20. Jha, B.K., Vatsal, V.H., Singh, T.P.: Navigating the fractional calcium dynamics of orai mechanism in polar dimensions. Cell Biochem. Biophys. (2024). https://doi.org/10.1007/s12013-024-01462-y

21. Vatsal, V.H., Jha, B.K., Singh, T.P.: Generalised neuronal calcium dynamics of membrane and ER in the polar dimension. Cell Biochem. Biophys. (2024). https://doi.org/10.1007/s12013-024-01425-3
22. Vatsal, V.H., Jha, B.K., Singh, T.P.: Deciphering two-dimensional calcium fractional diffusion of membrane flux in neuron. J. Appl. Math. Comput. (2024). https://doi.org/10.1007/s12190-024-02115-2
23. Pawar, A., Pardasani, K.R.: Effect of disturbances in neuronal calcium and IP3 dynamics on β-amyloid production and degradation. Cogn Neurodyn. **17** (2023). https://doi.org/10.1007/s11571-022-09815-0
24. Pawar, A., Pardasani, K.R.: Nonlinear system dynamics of calcium and nitric oxide due to cell memory and superdiffusion in neurons. Commun. Theor. Phys. **76**, 055002 (2024). https://doi.org/10.1088/1572-9494/ad35b4
25. Keener, J., Sneyd, J. (eds.): Mathematical Physiology. Springer, New York, New York, NY (2009)
26. Gil, D., Guse, A.H., Dupont, G.: Three-dimensional model of sub-plasmalemmal Ca2+ microdomains evoked by the interplay between ORAI1 and InsP3 Receptors. Front Immunol. **12** (2021). https://doi.org/10.3389/fimmu.2021.659790
27. Podlubny, I.: Fractional differential equations : an introduction to fractional derivatives, fractional differential equations, to methods of their solution and some of their applications (1999)
28. Joshi, H., Jha, B.K.: 2D dynamic analysis of the disturbances in the calcium neuronal model and its implications in neurodegenerative disease. Cogn. Neurodyn. (2022). https://doi.org/10.1007/s11571-022-09903-1
29. Mainardi, F., Luchko, Y., Pagnini, G.: The fundamental solution of the space-time fractional diffusion equation (2007)
30. Joshi, H., Jha, B.K.: On a reaction–diffusion model for calcium dynamics in neurons with Mittag–Leffler memory. Eur Phys J Plus. **136**. https://doi.org/10.1140/epjp/s13360-021-01610-w
31. Kilbas, A., Srivastava, H., Trujillo, J.: Theory and applications of fractional differential equations (2006)
32. Bezprozvanny, I.: Calcium signaling and neurodegenerative diseases. Trends Mol Med. **15** (2009). https://doi.org/10.1016/j.molmed.2009.01.001

Numerical Study of Thermo-Convective Instability of Au–Fe_3O_4 Hybrid Bi-Viscous Bingham Nanofluid

Sanjalee Maheshwari and Rajesh Sharma

Abstract In the present work, the thermal convective instability of hybrid bi-viscous Bingham nanofluid is numerically investigated using the Chebyshev spectral method. Considering the importance of gold and iron oxide nanoparticles in medical fields for optimized and targeted drug delivery, blood suspended with gold and iron oxide nanoparticles is taken as the working fluid. Flow dynamics are modeled using the modified Brinkman model for hybrid fluid and bi-viscous Bingham fluid model under the effect of Brownian, thermophoretic diffusion, and internal heat generation. The eigenvalue problem is derived from the linear stability analysis and normal mode technique. Further, the resulting eigenvalue problem is solved using the Chebyshev spectral method. The effect of pertinent flow governing parameters on the threshold of onset of convection is graphically presented and discussed in detail. It is observed that bi-viscous Bingham fluid suspended with only iron oxide nanoparticles is highly stable in the absence of an internal heat source.

Keywords Bi-viscous Bingham fluid · Heat source · Hybrid nanofluid · Spectral method

1 Introduction

Choi [1] coined the term "nanoliquid" to describe a conventional fluid containing nanometer-sized metallic and non-metallic particles. Nanoliquids are widely utilized in various industries such as automotive, electronics, and medicine due to their improved heat transfer properties. Despite their broad application, there is a growing interest in developing more efficient liquids than nanoliquids. The concept of "hybrid

S. Maheshwari (✉)
School of Basic Sciences, IIIT Una, Shimla, Himachal Pradesh, India
e-mail: maheshwari15sanjlee@gmail.com

R. Sharma
Department of Mathematics, North Eastern Hill University, Shillong, Meghalaya, India
e-mail: raj.juit@gmail.com

R. K. Sharma et al. (eds.), *Mathematical Theory and Simulation of Scientific Problems*, Springer Proceedings in Mathematics & Statistics 487,
https://doi.org/10.1007/978-981-96-2579-6_15

nanoliquid" has emerged, which involves suspending different nanoparticles in a specific base liquid. Hybrid nanoliquids exhibit higher thermal conductivity compared to regular liquids and nanoliquids, making them valuable for biomedical and industrial applications including electronic cooling, atomic reactors, and drug delivery.

The superior heat conductivity of hybrid nanoliquids has attracted significant attention from researchers aiming to address real-world challenges. For instance, Devi and Devi [2] conducted a numerical study on the hydromagnetic flow of Cu–Al_2O_3 water-based hybrid nanoliquid past a permeable stretching plane. Hayat and Nadeem [3] examined the effects of the chemical reaction and heat generation on radiating hybrid nanoliquid flowing past a stretching surface, while Hayat et al. [4] explored the same problem with partial slip boundary conditions. Ashorynejad and Shahriari [5] numerically analyzed the natural convection of Al_2O_3–Cu/water hybrid nanoliquid in a wavy wall cavity, revealing that a stronger magnetic field slows down the heat transfer rate. Additionally, Waini et al. [6] obtained dual solutions for the steady flow of hybrid nanoliquid past a stretching plane, conducting temporal stability analysis to determine the stability of the solutions. Nadeem et al. [7] evaluated the heat transfer rate of Cu–Al_2O_3 hybrid nanoliquid in an exponentially stretching channel, while Wakif et al. [8] utilized a generalized Buongiorno's model to investigate thermo-magneto convective instability in hybrid nanoliquid.

The category of non-Newtonian fluids known as bi-viscous Bingham fluids holds significant importance. This classification encompasses liquids like blood, sauce, chocolate, and honey, which exhibit zero viscosity at infinite shear rates and vice versa. Chaturani and Samy [9] utilized the bi-viscous Bingham model to characterize blood rheology. Sarojamma and Vendabai [10] conducted an analytical study on the uniformly heated boundary layer flow of bi-viscous Bingham nanoliquid around a cylinder stretching radially. Several notable contributions to this field have been documented by [11–13]. Aman et al. [14] conducted numerical investigations on the flow of hybrid bi-viscous Bingham nanoliquid through a vertical channel, incorporating Caputo time-fractional derivatives.

In recent years, the utilization of nanoliquids has seen a significant rise in the treatment of various medical conditions. Specifically, gold nanoparticles are extensively employed in cancer therapy due to their large size and high absorption capacity, while iron oxide nanoparticles are commonly used for magnetic resonance imaging, magnetic hyperthermia, and drug delivery owing to their superparamagnetism and size [15]. The transfer of heat between the heart and the surrounding surface is notably influenced by nanoliquids through convective processes within the blood. Therefore, investigating the convective instability of blood in the presence of bi-viscous Bingham liquid containing gold and iron oxide nanoparticles, along with an internal heat source, holds paramount importance in the medical field for enhancing healthcare practices.

Based on the literature reviewed above, it is noteworthy that there has been no prior investigation into the convective instability of the Au–Fe_3O_4 hybrid bi-viscous Bingham nanofluid. Therefore, this article aims to examine the initiation of convective currents in bi-viscous Bingham fluid (blood) containing gold and iron oxide

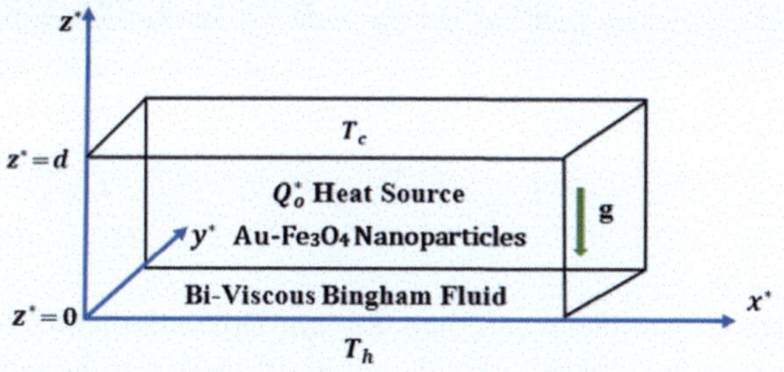

Fig. 1 Geometry and Physical representation of the problem

nanoparticles. The modeling of flow constitutive equations considers the presence of a non-uniform internal heat source.

2 Mathematical Formulation

Figure 1 depicts a horizontal enclosure with a finite thickness d and infinite length, initially containing a bi-viscous Bingham hybrid nanofluid. An internal heat source Q_o^*, varying across the system, is applied. The nanofluidic system experiences thermal disturbance by heating the bottom layer while maintaining the upper layer at a constant temperature T_c. To adhere to a realistic physical approach, a zero nanoparticle mass flux constraint is enforced at the boundaries. Flow constitutive equations for present stability problem [14, 16, 17] are

$$\nabla^* V^* = 0, \tag{1}$$

$$\rho_{hnf}\left(\frac{\partial}{\partial t^*} + (V^* \cdot \nabla^*)\right)V^* = -\nabla^* p^* + \mu_{hnf}\left(1 + \frac{1}{\gamma}\right)\nabla^{*2}V^* - \rho_{hnf}[1 - \beta_T(T^* - T_c) + \beta_C(\phi^* - \phi_h)]g\hat{e}_z, \tag{2}$$

$$(\rho C_p)_{hnf}\left(\frac{\partial}{\partial t^*} + (V^* \cdot \nabla^*)\right)T^* = k_{hnf}\nabla^{*2}T^* + \nabla^* T^* + Q_o{}^*(T^* - T_c) + (\rho C_p)_{np}\left(D_B\nabla^*\phi^* + \frac{D_T}{T_c}(\nabla^* T^*)\right), \tag{3}$$

$$\left(\frac{\partial}{\partial t^*} + (V^* \cdot \nabla^*)\right)\phi^* = D_B\nabla^{*2}\phi^* + \frac{D_T}{T_c}\nabla^{*2}T^*. \tag{4}$$

Based on the above assumption, the problem under consideration has the following periphery conditions:

$$w^* = 0,\ \frac{\partial w^*}{\partial z^*} = 0,\ T^* = T_h,\ D_B\frac{\partial \phi^*}{\partial z^*} + \frac{D_T}{T_c}\frac{\partial T^*}{\partial z^*} = 0 \quad \text{at } z^* = 0 \text{ (the lower wall).} \tag{5}$$

$$w^* = 0,\ \frac{\partial w^*}{\partial z^*} = 0,\ T^* = T_c,\ D_B\frac{\partial \phi^*}{\partial z^*} + \frac{D_T}{T_c}\frac{\partial T^*}{\partial z^*} = 0 \quad \text{at } z^* = d \text{ (the upper wall).} \tag{6}$$

Here coefficients D_B and D_T are Brownian and thermophoretic diffusion coefficients that characterize the arbitrary motion of nanoparticles in a hybrid nanofluid. T_h is temperature at the lower wall, γ is bi-viscous Bingham fluid parameter, β_T is thermal expansion coefficient, and β_C is mass expansion coefficient. Throughout the manuscript subscript hnf is used for hybrid nanofluid, $p1$ for iron oxide nanoparticles, $p2$ for gold nanoparticles, and f for base fluid blood.

The thermophysical characteristics such as viscosity (μ), density (ρ), thermal conductivity (k), and thermal expansion coefficient (β_C) of hybrid nanofluid are calculated using the expression given in the study [18] using the thermophysical properties of base fluid blood, gold, and iron oxide nanoparticles given in the same study. The following dimensionless variables are incorporated to make the constitutive PDEs non-dimensional:

$$(x, y, z) = \frac{x^*, y^*, z^*}{d},\ t = \frac{t^*\alpha_{hnf}}{d^2},\ T = \frac{T - T_c}{T_h - T_c},\ \phi = \frac{\phi^* - \phi_h}{\phi_h},\ (u, v, w) = \frac{(u^*, v^*, w^*)d}{\alpha_{hnf}},$$
$$p = \frac{p^* d^2}{\mu_{hnf}\alpha_{hnf}},\ \alpha_{hnf} = \frac{k_{hnf}}{(\rho C_p)_{hnf}},\ \tau = \frac{(\rho C_p)_{np}}{(\rho C_p)_{hnf}}. \tag{7}$$

The transformed dimensionless Eqs. 1–4 are given as follows:

$$\nabla \cdot V = 0, \tag{8}$$

$$\frac{1}{Pr}\left(\frac{\partial}{\partial t} + (V\cdot\nabla)\right)V = -\nabla p + \left(1 + \frac{1}{\gamma}\right)\nabla^2 V + RaT\hat{e}_z - Rn\phi\hat{e}_z, \tag{9}$$

$$\left(\frac{\partial}{\partial t} + (V\cdot\nabla)\right)T = \nabla^2 T + \frac{Nb}{Le}\nabla\phi\nabla T + H_s T + \frac{NaNb}{Le}(\nabla T)^2, \tag{10}$$

$$\left(\frac{\partial}{\partial t} + (V\cdot\nabla)\right)\phi = \frac{1}{Le}\nabla^2\phi + \frac{Na}{Le}\nabla^2 T. \tag{11}$$

And, the dimensionless form of boundary condition is given as follows:

$$w = 0,\ \frac{\partial w}{\partial z} = 0,\ T = 1,\ \frac{\partial \phi}{\partial z} + Na\frac{\partial T}{\partial z} = 0 \quad \text{at } z = 0 \text{ (the lower wall).} \tag{12}$$

$$w = 0,\ \frac{\partial w}{\partial z} = 0,\ T = 0,\ \frac{\partial \phi}{\partial z} + Na\frac{\partial T}{\partial z} = 0 \quad \text{at } z = 1 \text{ (the upper wall).} \tag{13}$$

The expressions for dimensionless physical parameters appearing in Eqs. 8–11 are

$$Ra = \frac{g(\rho\beta)_{hnf}(T_h - T_c)d^3}{\mu_{hnf}\alpha_{hnf}},\ Le = \frac{\alpha_{hnf}}{D_B},\ H_s = \frac{Q_o{}^* d^2}{k_{hnf}},\ Nb = \tau\phi_h,$$
$$Na = \frac{D_T(T_h - T_c)}{D_B\phi_h T_c},\ Rn = \frac{(\rho_{np} - \rho_f)g\phi_h d^3}{\mu_{hnf}\alpha_{hnf}}. \tag{14}$$

Here Ra is the thermal Rayleigh number, Le is the Lewis number, Hs is the non-dimensional heat source parameter, Nb is particle density increment, Na is the modified diffusivity ratio, and Rn is the nanoparticle concentration Rayleigh number.

3 Primary Flow and Linear Stability Analysis

Initially, the hybrid nanofluidic system is considered to be in a quiescent state, and physical properties such as pressure, temperature, and nanoparticle volume fraction are considered to be the only function of z, expressed as follows:

$$V = (0, 0, 0),\ p = \tilde{p}(z),\ T = \tilde{T}(z),\ \phi = \tilde{\phi}(z). \tag{15}$$

After incorporating Eq. 15, in Eqs.8–11 and using the boundary conditions, the following solutions for primary flow are obtained:

$$\tilde{T}(z) = \frac{Sin(1 - z)\sqrt{H_s}}{Sin\sqrt{H_s}}, \tag{16}$$

$$\frac{d\tilde{\phi}}{dz} = -Na\frac{d\tilde{T}}{dz}. \tag{17}$$

To test the stability of the primary flow, small disturbances are superimposed on the primary state, in the following way:

$$V = V',\ p = \tilde{p} + p',\ T = \tilde{T} + T',\ \phi = \tilde{\phi} + \phi'. \tag{18}$$

After imposing the condition given by Eq. 18 into Eqs. 8–11 and taking only linear terms into account, the following equations involving perturbed quantities are obtained:

$$\nabla \cdot V' = 0, \tag{19}$$

$$\frac{1}{Pr}\frac{\partial V'}{\partial t} = -\nabla p' + \left(1 + \frac{1}{\gamma}\right)\nabla^2 V' + RaT'\hat{e}_z - Rn\phi'\hat{e}_z, \tag{20}$$

$$\frac{\partial T'}{\partial t} + w'\frac{d\tilde{T}}{dz} = (\nabla^2 + H_s)T' + \frac{Nb}{Le}\frac{\partial \tilde{T}}{\partial z}\nabla T' + \frac{Nb}{Le}\frac{\partial \tilde{T}}{\partial z}\nabla\phi', \tag{21}$$

$$\frac{\partial \phi'}{\partial t} - Na\frac{d\tilde{T}}{dz}w' = \frac{1}{Le}\nabla^2\phi' + \frac{Na}{Le}\nabla^2 T'. \tag{22}$$

Boundary conditions subjected to the above-perturbed equations are

$$w^{'} = 0, \ \frac{\partial w^{'}}{\partial z} = 0, \ T^{'} = 0, \ \frac{\partial \phi^{'}}{\partial z} + Na\frac{\partial T^{'}}{\partial z} = 0, \quad \text{at } z = 0 \ \& \ 1. \tag{23}$$

Now applying the operator *curl twice* on both sides of Eq. 20, and using the identity $curl\ curl = grad\ div - grad^2$. z-component perturbed equations are given as

$$\frac{1}{Pr}\frac{\partial}{\partial t}(\nabla^2 w^{'}) = \left(1 + \frac{1}{\gamma}\right)\nabla^4 w^{'} + Ra\nabla_h^2 T^{'} - Rn\nabla_h^2 \phi^{'}, \tag{24}$$

here $\nabla_h^2 = \left(\frac{\partial^2}{\partial x^2} + \frac{\partial^2}{\partial y^2}\right)$.

4 Normal Mode Analysis

Due to the translation invariance of constitutive equations in x- and y-direction, perturbed quantities $w^{'}$, $T^{'}$, and $\phi^{'}$ are considered as two-dimensional periodic time-dependent waves and can be written as follows:

$$\begin{bmatrix} w^{'}(x, y, z, t) \\ T^{'}(x, y, z, t) \\ \phi^{'}(x, y, z, t) \end{bmatrix} = \begin{bmatrix} W(z) \\ \Theta(z) \\ \Phi(z) \end{bmatrix} e^{i(a_x + a_y) + \lambda t}. \tag{25}$$

In the present stability problem, the oscillatory mode of convection has been completely ruled out due to the absence of two contrary buoyancy forces that are responsible for oscillatory convection. Therefore only stationary convection mode ($\lambda = 0$) is possible. After, incorporating Eq. 25 in Eqs. 21 and 22, we get the following non-dimensional stability equations:

$$\left(1 + \frac{1}{\gamma}\right)(D^2 - a^2)^2 W - Raa^2\theta + Rna^2\Phi = 0, \tag{26}$$

$$(D\tilde{T})W - \left(D^2 - a^2 + H_s + \frac{NaNb}{Le}D\tilde{T}D\right)\theta - \frac{Nb}{Le}D\tilde{T}D\Phi = 0, \tag{27}$$

$$Na(D\tilde{T})W + \frac{Na}{Le}(D^2 - a^2)\theta + \frac{1}{Le}(D^2 - a^2)\Phi = 0. \tag{28}$$

Here, a_x and a_y are the wave numbers in x- and y-directions, respectively, such that $a_x^2 + a_y^2 = a^2$ and $\frac{d}{dz} \equiv D$. The above equations are solved by using the following boundary conditions:

$$W = 0, \ DW = 0, \ \theta = 0, \ D\Phi + NaD\theta = 0, \quad \text{at } z = 0 \ \& \ 1. \tag{29}$$

Now, to convert the Neumann boundary condition into the Dirichlet one, we introduce the following transformation proposed by [19]:

$$\varGamma = -D\Phi - NaD\Theta.$$

After applying the transformation and putting the value of primary flow quantities, the transformed form of Eqs. 26–28 is given as

$$\left(\left(1+\frac{1}{\gamma}\right)(D^2-a^2)^2 + RnNaLeD\tilde{T}\right)W - Raa^2\theta - RnNaa^2\theta - RnD\varGamma = 0, \quad (30)$$

$$D\tilde{T}W - (D^2 - a^2 + H_s)\theta + \frac{Nb}{Le}D\tilde{T}\,\varGamma = 0, \quad (31)$$

$$Na(D\tilde{T}D + D^2\tilde{T})W - \frac{1}{Le}(D^2 - a^2)\varGamma = 0. \quad (32)$$

The above simplified differential equations will be solved together with the following boundary conditions:

$$W = 0,\ DW = 0,\ \theta = 0,\ \varGamma = 0, \qquad \text{at } z = 0\ \&\ 1. \quad (33)$$

5 Solution Methodology

To study the stability of the present problem, the Chebyshev–Gauss–Lobatto spectral method (CGLSM) is employed. The system of Eqs. 30–32 constitutes an eigenvalue problem along with the boundary condition given by Eq. 33. The CGLSM is employed as described in the study [20]. The method has been validated by comparing the results obtained for the regular fluid with the earlier published works and found to be in good agreement as presented in Table 1.

Table 1 Comparison of the present results with the results of [21, 22], for regular fluid

	Yadav et al.		Shivkumar and Suma		Present results	
H_s	Ra_c	a_c	Ra_c	a_c	Ra_c	a_c
1	1704.5239	3.119	1704.5264	3.1189	1704.5264	3.119
2	1694.9479	3.127	1694.9501	3.3.1265	1694.9501	3.126
10	1462.8682	3.304	1462.8609	3.3036	1462.8609	3.304
20	1118.4591	3.529	1118.4301	3.5291	1118.4301	3.529

6 Results and Discussion

In this section, the influence of various non-dimensional flow-characterizing parameters on the threshold of convection in the hybrid nanofluid layer confined between two rigid channel walls has been graphically discussed. To understand the thermal stability of the fluidic system, neutral stability curves are sketched between the thermal Rayleigh number and wavenumber by taking the thermal Rayleigh number as a function of the wavenumber and various flow-characterizing parameters. The region between the curve and horizontal axis (representing the wavenumber) is considered a stable region. Hence, the vertical upward shift of the curve signifies the delay in the onset of convection. Further, the value of the wavenumber (critical wavenumber) corresponding to which the minimum value of the thermal Rayleigh number (critical thermal Rayleigh number) occurs helps to calculate the size of convection cells. Since convection cell size is calculated by the expression $\frac{2\pi}{a_c}$, the left horizontal shift of the minimum value of thermal Rayleigh number refers to the increase in the size of convection cells.

Figure 2 is delineated to study the influence of the bi-viscous Bingham fluid parameter on the threshold of onset of convection. The infinite value of the parameter refers to the Newtonian fluid and from the figure it can be observed that with the increase in the bi-viscous Bingham fluid parameter, the critical value of the thermal Rayleigh number decreases. This observation is physically valid due to the fact of direct correspondence between fluid parameters and fluid viscosity. Further, it can also be concluded that the use of non-Newtonian bi-viscous Bingham base fluid more stabi-

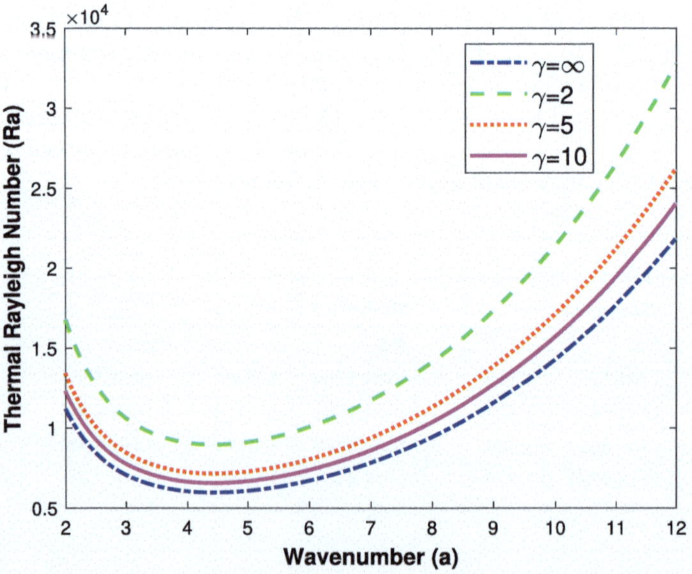

Fig. 2 Plots between Ra and a for various values of bi-viscous Bingham fluid parameter

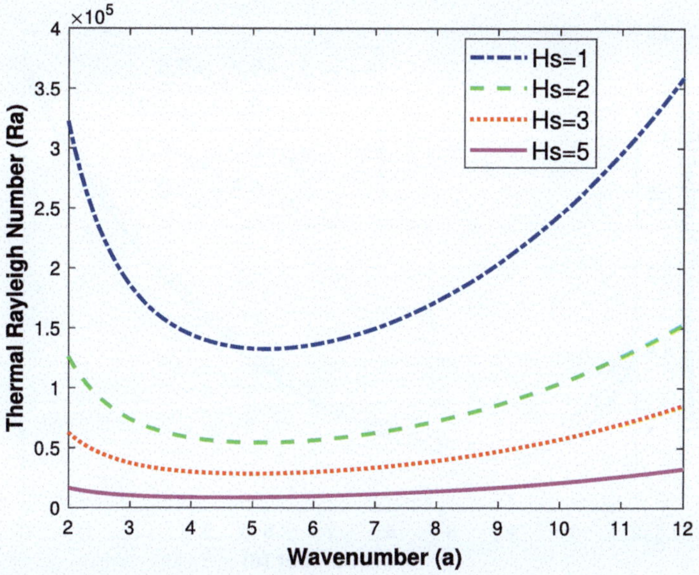

Fig. 3 Plots between Ra and a for various values of heat source parameter

lizes the system than that of the regular Newtonian base fluid. This outcome signifies the importance of considering the non-Newtonian behavior blood during the targeted drug administration. Since the critical value of the thermal Rayleigh number occurs at lower values of critical wavenumber with increasing value of bi-viscous Bingham fluid parameter, the size of the convection cell increases.

In Fig. 3, the effect of the heat source parameter on the threshold of onset of convection is shown. Advancement in the onset of convection with the strengthening of heat source parameters is observed. This outcome is anticipated since the advancement of the heat source parameter corresponds to the more heat supply in the system, which causes more thermal disturbances and hence results in the prior onset of convection. The effect of nanoparticle volume fraction on the onset of convection is shown in Fig. 4. From the plots, the system is found to be more stable when only

Table 2 Numerical values of critical thermal Rayleigh number for various values of concentration Rayleigh number and Lewis number

Rn	Le	Ra_c	a_c
1	2	9002.9	4.40
2	2	9007.9	4.40
5	2	9022.8	4.40
8	2	9037.8	4.40
2	1	9000.9	4.40
2	3	9014.8	4.40
2	5	9028.8	4.40

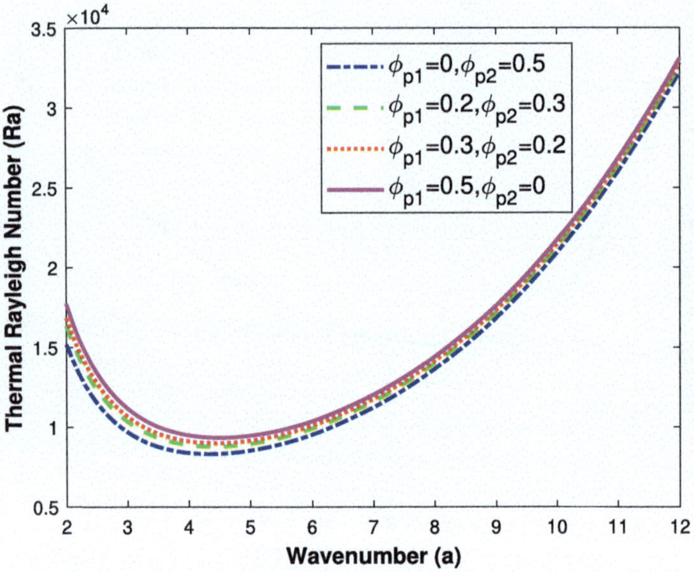

Fig. 4 Plots between Ra and a for various values of nanoparticle concentration

iron oxide nanoparticles are present ($\phi_{p1} = 0$ and $\phi_{p2} = 0.5$) in the system due to the relatively lighter iron oxide nanoparticles than the gold nanoparticles.

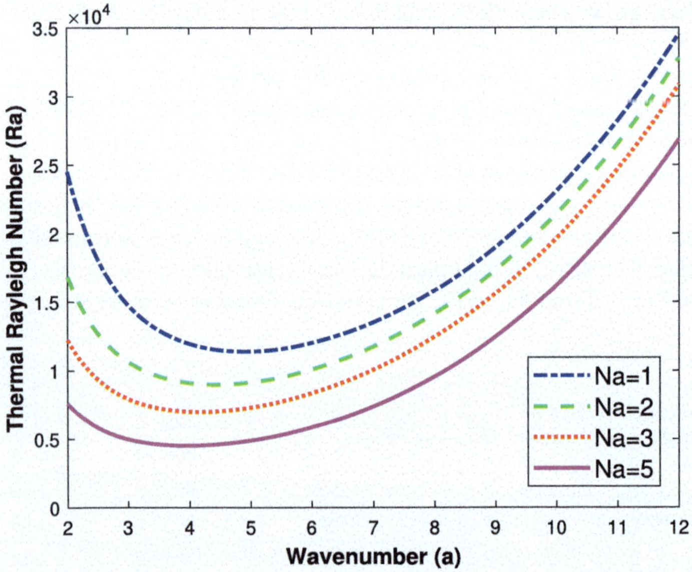

Fig. 5 Plots between Ra and a for various values of modified diffusivity parameter

The modified diffusivity ratio is directly proportional to the ratio thermophoretic to Brownian diffusion and temperature difference between the channel walls. An increase in the modified diffusivity ratio signifies more diffusion through the temperature difference than due to the Brownian motion. Therefore, with the increase in the modified diffusivity ratio thermal stability of the system decreases but the size of the convection cell increases as shown in Fig. 5. The influence of concentration Rayleigh number and Lewis number of the critical value of thermal Rayleigh number is numerically computed and presented in Table 2. It is observed that with the increase in the concentration of Rayleigh number for fixed values of Lewis number, critical thermal Rayleigh number increases. A similar effect is observed for increasing the value of the Lewis number. Hence increase in both the concentration Rayleigh number and Lewis number stabilizes the system and delays the onset of convection in the fluid layer.

7 Conclusion

Thermal stability of the internally heated hybrid non-Newtonian nanofluid has been numerically discussed using the linear stability theory, normal mode technique, and Chebyshev–Gauss–Lobatto spectral method. Base fluid suspended with gold and iron oxide nanoparticles is considered as the working fluid. The accuracy of the used method is proved by comparing the outcomes of the limiting cases of the study with the previously published works. The key findings of the study are abridged as follows:

- Bi-viscous Bingham non-Newtonian base fluid stabilizes the system and delays the starting of convective current in the fluidic system.
- Heat supply thermally distraughts the system and advances the onset of convection.
- The presence of heavier gold nanoparticles sets the convection earlier and hence destabilizes the system.
- The nanofluidic system will be most stable when suspended with iron oxide nanoparticles and the system is kept in the absence of an internal heat source.

References

1. Choi, S.: Enhancing thermal conductivity of fluids with nanoparticles. ASME Publ. Fed. **231**, 99–106 (1995)
2. Devi, S., Devi, S.S.U.: Numerical investigation of hydromagnetic hybrid Cu–Al_2O_3/H_2O nanofluid flow over a permeable stretching sheet with suction. Int. J. Nonlinear Sci. Numer. Simul. **17**, 249–257
3. Hayat, T., Nadeem, S.: Heat transfer enhancement with Ag-CuO/water hybrid nanofluid. Results Phys. **7**, 2317–2324 (2017)
4. Hayat, T., Nadeem, S., Khan, A.U.: Rotating flow of Ag–CuO/H_2O hybrid nanofluid with radiation and partial slip boundary effects. Eur. Phys. J. E **41**, 75 (2018)
5. Ashorynejad, H., Shahriari, A.: MHD natural convection of hybrid nanofluid in an open wavy cavity. Results Phys. **9**, 440–455 (2018)
6. Waini, I., Ishak, A., Pop, I.: Unsteady flow and heat transfer past a stretching/shrinking sheet in a hybrid nanofluid. Int. J. Heat Mass Transf. **136**, 288–297 (2019)

7. Nadeem, S., Abbas, N., Malik, M.Y.: Inspection of hybrid based nanofluid flow over a curved surface. Comput. Methods Programs Biomed. **189**, 105193 (2020)
8. Wakif, A., Chamkha, A., Thumma, T., Animasaun, I., Sehaqui, R.: Thermal radiation and surface roughness effects on the thermo-magneto-hydrodynamic stability of alumina-copper oxide hybrid nanofluids utilizing the generalized Buongiorno's nanofluid model. J. Thermal Anal. Calorim. **141**, 1201–1220 (2020)
9. Chaturani, P., Samy, R.P.: Pulsatile flow of Casson's fluid through stenosed arteries with applications to blood flow. Biorheology **23**, 499–511 (1986)
10. Sarojamma, G., Vendabai, K.: Boundary layer flow of a Casson nanofluid past a vertical exponentially stretching cylinder in the presence of a transverse magnetic field with internal heat generation/absorption. Int. J. Math. Comput. Sci. **9**, 138–143 (2015)
11. Naqvi, S., Muhammad, T., Asma, M.: Hydromagnetic flow of Casson nanofluid over a porous stretching cylinder with Newtonian heat and mass conditions. Phys. A Stat. Mech. Appl. **550**, 123988 (2020)
12. Aneja, M., Chandra, A., Sharma, S.: Natural convection in a partially heated porous cavity to Casson fluid. Int. Commun. Heat Mass Transf. **114**, 104555 (2020)
13. Trivedi, M., Otegbeye, O., Ansari, M., Motsa, S.S.: A paired quasi-linearization on magneto-hydrodynamic flow and heat transfer of Casson nanofluid with Hall effects. J. Appl. Comput. Mech. **5**, 849–860 (2019)
14. Aman, S., Zokri, S., Ismail, Z., Salleh, M., Khan, I.: Casson model of MHD flow of SA-based hybrid nanofluid using Caputo Time-Fractional models. Defect Diffus. Forum. **390**, 83–90 (2019)
15. Jahangirian, H., Kalantari, K., Izadiyan, Z., Rafiee-Moghaddam, R., Shameli, K., Webster, T.: A review of small molecules and drug delivery applications using gold and iron nanoparticles. Int. J. Nanomed. **14**, 1633 (2019)
16. Wakif, A., Boulahia, Z., Mishra, S., Rashidi, M., Sehaqui, R.: Influence of a uniform transverse magnetic field on the thermo-hydrodynamic stability in water-based nanofluids with metallic nanoparticles using the generalized Buongiorno's mathematical model. Eur. Phys. J. Plus. **133**, 181 (2018)
17. Yadav, D., Wang, J.: Convective heat transport in a heat generating porous layer saturated by a non-newtonian nanofluid. Heat Transf. Eng. **40**, 1363–1382 (2019)
18. Maheshwari, S., Sharma, Y., Yadav, O.: Convective instability of blood-based Au–Fe_3O_4 hybrid nanoliquid under the presence of magnetic field with internal heat source: application to cancer treatment. ZAMM J. Appl. Math. Mech./Zeitschrift Für Angewandte Mathematik Und Mechanik. **103**, e202200518 (2023)
19. Wakif, A., Boulahia, Z., Amine, A., Animasaun, I., Afridi, M., Qasimd, M., Sehaqui, R.: Magneto-convection of alumina-water nanofluid within thin horizontal layers using the revised generalized Buongiorno's model. Front. Heat Mass Transf. (FHMT) **12**, 1–15 (2018)
20. Sanjalee, Sharma, R.: Analysis of magneto-thermo-bioconvection of nanofluid containing gyrotactic microorganisms through porous media. J. Nanofluids **11**, 979–987 (2022)
21. Yadav, D., Bhargava, R., Agrawal, G.: Boundary and internal heat source effects on the onset of Darcy-Brinkman convection in a porous layer saturated by nanofluid. Int. J. Therm. Sci. **60**, 244–254 (2012)
22. Shivakumara, I., Suma, S.: Effects of throughflow and internal heat generation on the onset of convection in a fluid layer. Acta Mech. **140**, 207–217 (2000)

The manufacturer's authorised representative in the EU is Springer Nature Customer Service Centre GmbH, Europaplatz 3, 69115 Heidelberg, Germany. If you have any concerns regarding our products, please contact ProductSafety@springernature.com

Printed and bound by CPI Group (UK) Ltd, Croydon, CR0 4YY
01/10/2026
02225935-0001